AF744729

Computer-aided Nonlinear Control System Design

Amir Nassirharand

Computer-aided Nonlinear Control System Design

Using Describing Function Models

Springer

Amir Nassirharand
Faculty of Engineering
The University of Nottingham
Malaysia Campus
Jalan Broga
43500 Semenyih
Malaysia

ISBN 978-1-4471-2148-0 e-ISBN 978-1-4471-2149-7
DOI 10.1007/978-1-4471-2149-7
Springer London Dordrecht Heidelberg New York

British Library Cataloguing in Publication Data
A catalogue record for this book is available from the British Library

Library of Congress Control Number: 2011942925

© Springer-Verlag London Limited 2012
Apart from any fair dealing for the purposes of research or private study, or criticism or review, as permitted under the Copyright, Designs and Patents Act 1988, this publication may only be reproduced, stored or transmitted, in any form or by any means, with the prior permission in writing of the publishers, or in the case of reprographic reproduction in accordance with the terms of licenses issued by the Copyright Licensing Agency. Enquiries concerning reproduction outside those terms should be sent to the publishers.
The use of registered names, trademarks, etc., in this publication does not imply, even in the absence of a specific statement, that such names are exempt from the relevant laws and regulations and therefore free for general use.
The publisher makes no representation, express or implied, with regard to the accuracy of the information contained in this book and cannot accept any legal responsibility or liability for any errors or omissions that may be made.

Printed on acid-free paper

Springer is part of Springer Science+Business Media (www.springer.com)

To Shabnam, Sam, and Daniel who suffer as I do my research
Of course to Nicole, Jessie, Kaydence, and Kiera
To my mother, Shokat Vahdat, and my late father, Hassan, who are proud of this work

Preface

The aim of this book is to give an in-depth treatment of computer-aided design of nonlinear control systems using describing function models. The approach is systematic, and it is based on a describing function platform. A large class of nonlinear systems is covered with no restrictions on the number, type, and arrangement of nonlinear terms, system order, and the number of inputs and/or the outputs. Unlike other books in the area of nonlinear control that are highly mathematically oriented, this book is software oriented.

It is believed that today's tight control system performance requirements do not allow a *linear* frame of mind for the designer. The designer's first choice to linearize a nonlinear model places so many unrealistic restrictions on the real model that drives up the costs of final product.

A full nonlinear frame of mind for design of nonlinear feedback control systems requires a computer-aided design approach. Many experts argue that such an approach would not guarantee *stability*. Here, without loss of generality, stability is demonstrated using the fact that if the sinusoidal input describing function models of a nonlinear system is obtainable, then that nonlinear system is stable around the operating regimes of interest.

I like to emphasize that having a full nonlinear frame of mind does not mean that designers should throw away other available linear and more restricted nonlinear control literature. On the contrary, one should capitalize on others' intellectual developments to bring good low-cost products for mankind. This book is a vehicle to account for nonlinear effects, and it simultaneously enables the designer to interface with other control system design works.

A person with a thorough knowledge of calculus and an elementary knowledge of numerical analysis and software engineering would most benefit from this book. This book should be of interest to aerospace, chemical, electrical, mechanical, and mechatronics students and practicing control engineers as well as other practical engineers and managers who are simply interested in design of low-cost real-life nonlinear feedback systems.

Many key MATLAB® functions are developed and presented to allow easy design of nonlinear feedback systems. The algorithms and the corresponding MATLAB® functions are demonstrated by solving example problems of the sort encountered in mechatronics and aerospace. A versatile controller design software based on the presented describing function design platform (using MATLAB® Graphics User Interface (GUI)) is under development which will be commercially available by early 2013.

Semenyih, Malaysia Amir Nassirharand

Acknowledgments

First, I like to sincerely thank my mentor, teacher, friend, counselor, adviser, and academic father, Professor James H. Taylor, for his continued support and encouragement throughout my academic and professional life. He has always been there to answer my questions.

I also like to thank my students. Special thanks to Professor Abhijit Patwardhan whose contributions on search of the parameter in simultaneous stabilization theory shed light on this important problem while he was pursuing his Master's degree at the University of Kentucky; my apologies if I was too hard on you! Mr. Li See Chew, my Ph.D. student, who is working on extension of some of the works presented here to the unstable case in a direct way and who is a pioneer in experimental work among my group members, deserves special thanks for putting up with my supervision. I sincerely thank Mr. Sze Hong Teh (University of Nottingham – Malaysia Campus, (UNMC)) for extending my Ph.D. work on dual-range linear controller design from the single-variable nonlinear case to the multi-variable nonlinear case. His other works on idle speed control of automobile engines, bank-angle control of aerospace vehicles with hard nonlinearities, and initiation of projects in chemical engineering and continuing to play with the spacecraft attitude control problem have contributed to preparation of this monograph. His collaboration in developing a versatile software package to support this monograph is acknowledged. I certainly have been lucky to have the opportunity to work with you. Support of Brandon (my Master's degree student at UNMC) is appreciated; I appreciate his aggressiveness to do his thesis on application of describing function techniques to nonlinear multivariable problem of satellite attitude control. Finally many thanks to all those undergraduate Mechanical Engineering and Mechatronics students at UNMC (Ahmed, Ajitpal, Aimen, Arun, Athira, Awaii, Bartholomew, Bushra, Chan, Chin, Gift, Hammad, Khawja, Lee, Lim, Mohd, Muhammad, Prem, Saeid, Sameer, Sanz, Sean, Shola, Tan, Tee, Yuvanraj) who selected my projects in the dynamic systems and control field as their final year theses, final year design projects, and final year group development projects.

Many thanks to Dr. Karimi and his research group (Norouz, Seyed Mehdi, Seyed Reza, Ali, Davood, ...) at K.N.T. University of Technology who taught me the practical aspects of modeling, simulation, and control of engineering systems; they substantially contributed to my professional growth that is reflected in this monograph. Contributions of Seyed Reza in solving the linear and nonlinear H_∞ control problems and nonlinear lead-lag design problem given in this monograph are greatly appreciated. Contributions of Norouz in application of algebraic linear multivariable controller design approach of C. T. Chen to describing function-based single range controllers are also acknowledged.

I like to express my sincere appreciation to Mr. Hadjirasouliha (former President of the Farab Company) and other personnel (Ardavan, Ata, Farshid, Farzan, Massoud, Mehdi, Mohamad Reza, Pouria, Samira, ...) at Farab for supporting me and allowing me to do my research while I was doing my engineering tasks.

Special thanks to Professor Seamus Garvey and Professor Michael Wood of University of Nottingham (UK Campus) for their invaluable support of my teaching in terms of providing their lecture notes and powerpoint slides as well as answering my detailed questions on Systems Engineering, Design of Experiments, Dynamic Systems, Reliability, and Robustness; this helped me allocate the extra time required to do my research and complete this monograph.

I appreciate the help, guidance, and supervision of Dr. Stewart McWilliam and Andrew Spowage while at UNMC. Special thanks to Professor Mike Cloke (former Dean) and Professor Ian Pashby (CEO) for giving me the opportunity to continue my academic career at UNMC.

The support, guidance, direction, and critique of Professor James Rowland (University of Kansas) of my initial proposal for my Ph.D. studies, which led to my pursuit of controller design for nonlinear systems, are acknowledged. His undergraduate text on control systems inspired me to pursue and develop closed-form solutions for design of linear lead-lag, linear PID, and other classical linear controllers (that are given in the Appendices) in an algorithmic and systematic way.

I also wish to express my gratitude to Mr. Oliver Jackson (Springer-Verlag London) for keeping an open mind in making the decision to go ahead with publication of this monograph. His understanding and support for the need for a software-oriented approach to design of nonlinear feedback control systems is noteworthy. It is noted that without the help and support of Ms. Charlotte Cross (Springer-Verlag London), this manuscript would not have been completed in time; special thanks to Ms. Cross for her help in obtaining the required permissions for use of copy righted material.

Last but not least, supports of my sisters Zohreh and Gita are appreciated. I also thank Mr. Hassan Bojnordiazad and Nasrin Rayat as they have touched my life in a positive way. Their encouragement keeps me going. I like to thank John and Patsy Hiser for contributing to my spiritual growth.

Thank you all!

Contents

Chapter 1
Introduction

1.1 What's New?

This is the first monograph on Computer-Aided Control System Design (CACSD) of nonlinear systems based on the describing function design platform that was originally proposed by Taylor in 1983. Since then, this platform has been developing as reported in the open literature. Other key characteristics that make this a new monograph follow:

- The considered class of nonlinear systems is of the following form:

$$\dot{x}(t) = f\,(x, u, t) \tag{1.1}$$

$$y(t) = g\,(x, u, t) \tag{1.2}$$

  where x is the state vector of dimension n, u is the input vector of dimension m, y is the system output vector of dimension l, and t is the time variable. There are no restrictions on the number, arrangement (e.g., could be in forward or feedback paths), and type of nonlinear terms (e.g., could be continuous or discontinuous), system order, and the number of inputs and/or the outputs (e.g., could be square or rectangular).
- Controller design for the considered class of nonlinear systems is describing-function-based.
- All design methods are presented in an easy to follow algorithmic format.

1.2 Design Platform

The design platform is sinusoidal-input describing-function-based, and the structure of the final closed-loop system is depicted in Fig. 1.1.

A. Nassirharand, *Computer-aided Nonlinear Control System Design: Using Describing Function Models*, DOI 10.1007/978-1-4471-2149-7_1,
© Springer-Verlag London Limited 2012

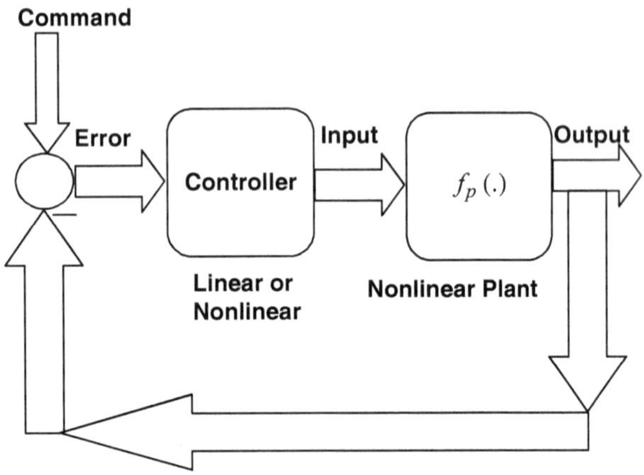

Fig. 1.1 The structure of the considered nonlinear feedback system

Describing functions are used for frequency domain modeling. Then, depending on the number of frequency domain models to be considered, a linear controller or a nonlinear controller design approach is outlined. In summary, the design platform considers three different cases: (1) single-range linear controller design (SRLCD), (2) dual-range linear controller design (DRLCD), and (3) multi-range nonlinear controller design (MRNCD).

When a designer starts designing a controller for a nonlinear system, he or she is not sure if a linear controller would suffice or a nonlinear controller should be designed. For this reason, it is recommended that designer should consider SRLCD, DRLCD, and MRNCD in that order. In other words, at first SRLCD is tried; if satisfactory results are not obtained, DRLCD should be executed. If DRLCD fails, MRNCD is tried.

With reference to items noted below (Taylor 1983; Taylor and Strobel 1985; Taylor and O'Donnell 1990; Nassirharand and Karimi 2006; Teh and Nassirharand 2010a), the following fact may be concluded. Sinusoidal-input describing function (SIDF) models are the best of the kind of models that characterize the input-output behavior of a nonlinear plant.

- Small-signal models, in contrast to the nonlinear plant, do not depend on amplitude level of the excitation signal.
- Random input or other types of describing function models are not able to characterize the dependency of the nonlinear plant on various frequencies of interest.
- Design based on one or more SIDF models, which covers the range of excitation amplitude levels, provides a solid basis for a robust design. This is because the sensitivity of the plant behavior to the amplitude level of the excitation signal is an important consideration in design of robust nonlinear feedback systems. The input-output SIDF models, without sacrificing the performance of the feedback system, may be utilized to assure robust system stability.

- One parameter is used in characterizing the input-output SIDF models; this parameter is the amplitude level of the excitation signal. This method of model parameterization is in contrast to that of linearizing methods. In linearizing methods, for each nonlinear term, one parameter, whose value is determined by taking the derivative of the nonlinear term, is used; however, amplitude of excitation signal characterizes each SIDF model. It should be noted that robust design methods for nonlinear systems based on small-signal linear models become very difficult to apply. In this case, the end result of the design is a feedback system that reduces system performance in order to achieve absolute stability.
- The controller synthesis method is applicable to nonlinear systems whose input-output models in the frequency domain are sensitive to the amplitude level of the excitation signal.

The primary goal of the controller synthesis is to create a closed-loop system whose behavior is fairly insensitive to the amplitude level of the excitation signal. In other words, the closed-loop system behavior would be insensitive to large perturbations of the plant including uncertainties and those caused by disturbances.

1.3 Objectives

Limited efforts in the past have been undertaken to provide a versatile CACSD setting for nonlinear control system design. This is primarily due to the fact that a large body of nonlinear control literature considers special classes of nonlinear systems. For example, geometric approaches assume Jacobian information is available (Hunt et al. 1987). A good part of sliding mode control literature considers nonlinear feedback systems that must be in companion form (Utkin et al. 1999). Quantitative feedback theory (QFT) considers a similar class of nonlinear systems to that considered in this monograph; however, its original formulation results in high-order linear controllers (Horowitz 1976). Nonlinear controller design using QFT approach is still developing.

The main objective of this monograph is to enable control engineering students and/or practicing engineers to design robust nonlinear feedback systems using a newly developed CACSD platform. Considering the fact that commercially available control engineering software utilities such as MATLAB® and MATRIXx® have advanced control engineering education and practice, it is safe to assume control engineers feel comfortable solving problems using a personal computer.

Another goal of the monograph is to generate interest in advancing the presented concepts to enable design of controllers for a larger class of systems. For example, consideration of higher harmonic effects is a good future work.

1.4 Software

The required software for design of robust nonlinear feedback systems (in form of a MATLAB® Graphics User Interface) is under development. However, the available software that is used to solve the example problems is outlined throughout the monograph. The listings for some of the developed functions are also given.

1.5 Organization of the Monograph

In Chap. 2, describing function generation from state-variable models is discussed. Example problems for single-variable and multivariable cases are solved. In Chap. 3, an algorithm is given for design of single-range controllers using describing function models followed by algorithm description and numerical examples for both single-variable and multivariable cases. In Chap. 4, an algorithm is given for design of dual-range controllers using two describing function models followed by example problems for both single-variable and multivariable cases. In Chap. 5, multi-range nonlinear controllers are designed for both single-variable and multivariable nonlinear systems; demonstration example problems of the sort encountered in robotics and aerospace are given to see the typical results that may be achieved. Finally, in Chap. 6, an experimental study of the nonlinear controller synthesis procedure outlined in Chap. 5 is presented with application to an unstable single-variable inverted pendulum problem.

Appendix A gives the closed-form solutions to the design of linear classical controllers that are used to arrive at the nonlinear controllers described in Chap. 5. Appendix B outlines an algebraic procedure for design of single-variable and multivariable linear plants. Appendix C includes the solution to other example problems encountered in single-range controller design. Appendix D includes application of the dual-range linear controller design problem to linear systems with integrity and nonlinear systems; the problems associated with the application of simultaneous stabilization theory to nonlinear systems are discussed.

Chapter 2
Frequency Domain Modeling

2.1 Early Formulation

Here, a frequency domain model of a nonlinear system is defined to be the sinusoidal input describing function (SIDF) model of that nonlinear system. For linear systems, frequency domain models are only a function of excitation frequency. Generally speaking, a frequency domain model of a nonlinear system is a function of both amplitude and frequency of excitation (Gelb and Vander Velde 1968; Atherton 1975). This dependency on the amplitude of excitation is the most important characteristic of a nonlinear system that must be taken into account when designing robust closed-loop feedback systems.

The basic version of the SIDF modeling approach is based on the following assumptions:

1. One single separable nonlinear term in the process.
2. The nonlinear term is time invariant. In other words, autonomous nonlinear processes are considered.
3. The input to the nonlinearity term may be characterized by a sinusoid of the following form:

$$u(t) = A\sin(\omega t), \tag{2.1}$$

   where u is the input, A is the amplitude of excitation, ω is the frequency of excitation, and t is the time variable.
4. Only the fundamental component of the output signal may be considered. In other words, it is assumed that higher harmonic effects are negligible.
5. The nonlinearity is odd. This means that nonlinearity is symmetric about the origin; e.g., $f(-u) = -f(u)$.

With the above assumptions, the following approach may be used to obtain the frequency domain model of a nonlinear process (Slotine and Li 1991). From the

A. Nassirharand, *Computer-aided Nonlinear Control System Design: Using Describing Function Models*, DOI 10.1007/978-1-4471-2149-7_2,
© Springer-Verlag London Limited 2012

third assumption, assume a sinusoidal input. Then, approximate the output signal, $y(t)$, by Fourier series.

$$y(t) = \frac{a_0}{2} + \sum_{n=1}^{\infty} [a_n \cos(n\omega t) + b_n \sin(n\omega t)], \tag{2.2}$$

where

$$a_n = \frac{1}{\pi} \int_{-\pi}^{\pi} y(t) \cos(n\omega t)\, d(\omega t), \tag{2.3}$$

$$b_n = \frac{1}{\pi} \int_{-\pi}^{\pi} y(t) \sin(n\omega t)\, d(\omega t). \tag{2.4}$$

With reference to the fourth and fifth Assumptions, Eq. 2.2 becomes

$$y(t) \approx b_1 \sin(\omega t). \tag{2.5}$$

By definition, the SIDF model of a nonlinear process is the ratio of the fundamental component of the nonlinear process output by that of the input sinusoid.

$$N(A) = \frac{b_1 \sin(\omega t)}{A \sin(\omega t)} = \frac{b_1}{A}. \tag{2.6}$$

Therefore, with the basic SIDF assumptions, the SIDF model for static operations is not a function of frequency.

2.1.1 *Example: Limiter*

Consider the limiter nonlinearity shown in Fig. 2.1. The nonlinear element has two regions. In order to find the describing function model of this nonlinearity, first assume a sinusoidal input of the form $u(t) = A\sin(\omega t)$. Then, evaluate Fourier integrals. Since the nonlinearity is odd, it is concluded that $a_0 = 0, a_1 = 0$.

$$y(t) = b_1 \sin(\omega t), \tag{2.7}$$

where

$$b_1 = \frac{1}{\pi} \int_{-\pi}^{\pi} y(t) \sin(\omega t)\, d(\omega t). \tag{2.8}$$

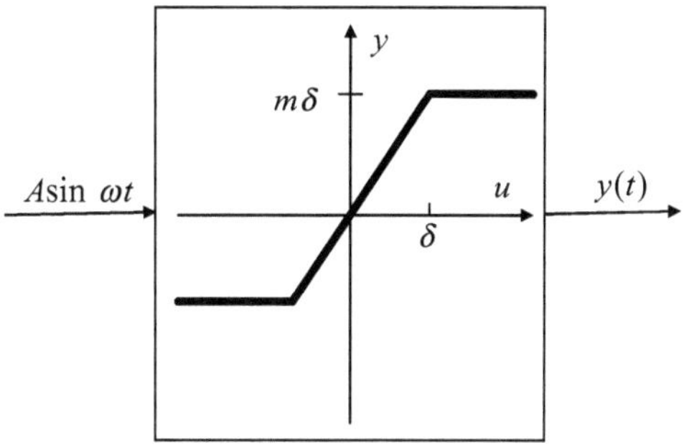

Fig. 2.1 The limiter nonlinearity

For $A \leq \delta$,

$$y(t) = mu(t) = mA\sin(\omega t), \tag{2.9}$$

From the expression for b_1

$$b_1 = \frac{mA}{\pi}\int_{-\pi}^{\pi} mA\sin(\omega t)\sin(\omega t)\,d(\omega t) = mA. \tag{2.10}$$

From Eq. 2.6,

$$N(A) = \frac{b_1}{A} = \frac{mA}{A} = m = N. \tag{2.11}$$

Therefore, for $u(t) \leq \delta$, the gain of a saturation is simply the constant m.
For $A > \delta$,

$$y(t) = \begin{cases} mA\sin(\omega t) & 0 \leq \omega t \leq \omega t_1 \\ m\delta & \omega t_1 \leq \omega t \leq \dfrac{\pi}{2} \end{cases}, \tag{2.12}$$

where $\omega t_1 = \sin^{-1}(\delta/A)$. Hence,

$$b_1 = \frac{1}{\pi}\int_{-\pi}^{\pi} y(t)\sin(\omega t)\,d(\omega t) = \frac{1}{\pi}\int_{0}^{2\pi} y(t)\sin(\omega t)\,d(\omega t)$$

$$= \frac{4}{\pi}\int_{0}^{\omega t_1} mA\sin^2(\omega t)\,d(\omega t) + \frac{4}{\pi}\int_{\omega t_1}^{\pi/2} m\delta\sin(\omega t)\,d(\omega t)$$

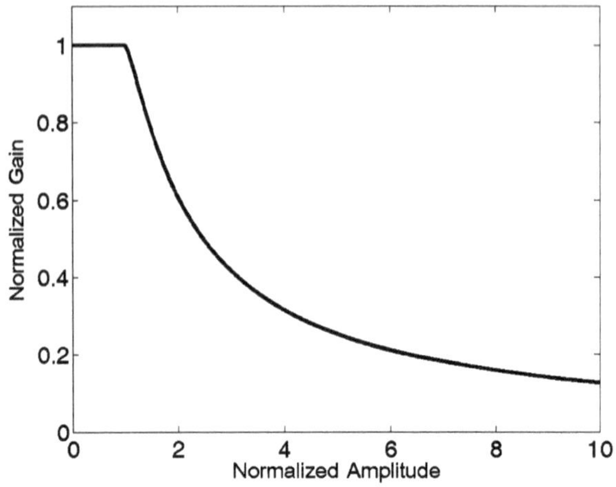

Fig. 2.2 The describing function model of the limiter

$$
\begin{aligned}
&= \frac{4mA}{\pi}\int_0^{\omega t_1}\frac{1}{2}\left[1-\cos\left(2\omega t\right)\right]d\left(\omega t\right) - \frac{4m\delta}{\pi}\cos\left(\omega t\right)\Big|_{\omega t_1}^{\pi/2}\\
&= \frac{2mA}{\pi}\left(\omega t\right)\Big|_0^{\omega t_1} - \frac{2mA}{\pi}\sin\left(2\omega t\right)\Big|_0^{\omega t_1} - \frac{4ma}{\pi}\left[0-\cos\left(\omega t_1\right)\right]\\
&= \frac{2mA}{\pi}\omega t_1 - \frac{mA}{\pi}2\sin\left(\omega t_1\right)\cos\left(\omega t_1\right) + \frac{4m\delta}{\pi}\cos\left(\omega t_1\right)\\
&= \frac{2mA}{\pi}\omega t_1 - \frac{2mA}{\pi}\frac{\delta}{A}\cos\left(\omega t_1\right) + \frac{4m\delta}{\pi}\cos\left(\omega t_1\right)\\
&= \frac{2mA}{\pi}\left[\omega t_1 + \frac{\delta}{A}\sqrt{1-\frac{\delta^2}{A^2}}\right].
\end{aligned}
\tag{2.13}
$$

Again, using Eq. 2.6 with $\omega t_1 = \sin^{-1}\left(\delta/A\right)$,

$$
N(A) = \frac{b_1}{A} = \frac{2mA}{\pi}\left[\sin^{-1}\left(\frac{\delta}{A}\right) + \frac{\delta}{A}\sqrt{1-\frac{\delta^2}{A^2}}\right]. \tag{2.14}
$$

The corresponding describing function plot is shown in Fig. 2.2. In this figure, the axis labeled by "normalized amplitude" corresponds to A/δ, and the axis labeled by "normalized gain" corresponds to $N(A)/m$ (Gelb and Vander Velde 1968).

Now, let us interpret the results. The saturation element has a constant gain over the linear region. This is shown by a horizontal line segment on the describing function graph (Fig. 2.2). However, inside the saturation nonlinearity, gain decreases as the amplitude of the excitation increases. Note that since gain multiplied by input must equal a constant, gain must decrease as the amplitude of the excitation increases. This behavior is shown in Fig. 2.2 by the curved region.

2.2 Modern Formulation

With the modern formulation (Taylor 1983; Taylor and Strobel 1985; Nassirharand 1987), the restrictions of the earlier formulation would be lifted. Also, the usual SIDF modeling limitations (such as the requirement that the input to each nonlinearity must be sinusoidal) do not exist with the modern formulation. The approach is based on simulation and evaluation of Fourier integrals. In majority of cases, the mathematical model of a system would be required before the task of controller design could begin. Even in those cases that a mathematical model is not required for controller design purposes, a mathematical model would be required to verify or validate design. However, with the presented design platform, if a mathematical model is not available, then experimental results may be used (Li and Nassirharand 2010).

In reality, the implemented mathematical models of real-life systems are hardly represented in an *academic* form. For example, the model would have *if* ... *then* ... type statements that do not usually fall into an academic class. In other words, most practical models are in terms of a computer code or a simulation model. In general, most of the models may be represented in the general framework given by Eqs. 1.1 and 1.2.

The computer models comprised of 5,000 lines of code are not unusual (Karimi et al. 2003). Once the computer model is developed, one faces a question. This question is how would one put this model in a form that available elegant controller design techniques would apply? The first easy way out is to obtain a small-signal model for the developed nonlinear model. In many cases, a linear model of a real-life system does not exist (e.g., see the robotic system studied in Sect. 2.2.1.2 (Nassirharand 1987)). This is primarily because real systems include multivalued and/or discontinuous nonlinearities whose derivatives do not exist; hence, a linear model (by replacing each nonlinear term with a term whose gain is the slope of the nonlinearity at the operating point) would not be obtainable.

One successful approach to put the complicated mathematical models (or nonlinear models whose small-signal models do not exist) in a form suitable for controller design purposes has been the describing function approach. The problem statement follows. Given the state variable model of a nonlinear system (see Eqs. 1.1 and 1.2), how would one characterize the input/output behavior of that system in the frequency domain?

2.2.1 Single-Input Single-Output Case

The stable plant is excited by a sinusoid of the following form:

$$u(t) = u_0 + a\cos(\omega t), \tag{2.15}$$

where u is the input to the plant, u_0 is the DC value of the input, a is the amplitude of the excitation, and ω is the frequency of excitation. The equations of motion are numerically integrated to obtain the output, $y(t)$. Once the output reaches steady state, the Fourier integrals for period k are evaluated; these integrals are of the following form:

$$I_{m,k} = \int_{(k-1)T}^{kT} y(t)e^{-jm\omega t}\mathrm{d}t, \tag{2.16}$$

where $k = 1, 2, 3, \ldots$ is the period index, $m = 0, 1, 2, \ldots$ is the harmonic index, and $T = 2\pi/\omega$ is the period. $I_{0,k}$ is the constant or DC component of the response, and the harmonic dependent transfer functions, $G_{m,k}$, are given by the following relation:

$$G_{m,k}(jm\omega; u_0, a) = \omega I_{m,k}/a\pi. \tag{2.17}$$

The Fourier integrals are assumed to have converged if the error in magnitude and phase of the computed pseudo frequency response in two consecutive cycles are *small.* The first harmonic SIDF models are given by $G_{1,k}$. The flow chart for this procedure is depicted in Fig. 2.3 (Nassirharand 1987).

2.2.1.1 Software

A MATLAB® function is available whose primary inputs are (1) the SIMULINK® file name corresponding to the model of the nonlinear plant, (2) an array of excitation amplitudes, and (3) an array of excitation frequencies. The primary outputs are the real and imaginary parts of the describing function models as a function of excitation amplitudes and frequencies. This software is based on an earlier work of Nassirharand in 1987.

2.2.1.2 Example: A Servo System

Consider the servo system shown in Fig. 2.4, which was originally considered by Taylor and Strobel 1985 and later by Nassirharand et al. 1988. The mathematical model of the nonlinear plant is given by Eqs. 2.18–2.21.

$$\dot{x}_1 = x_2 \tag{2.18}$$

$$\dot{x}_2 = T_{\mathrm{m}}/J \tag{2.19}$$

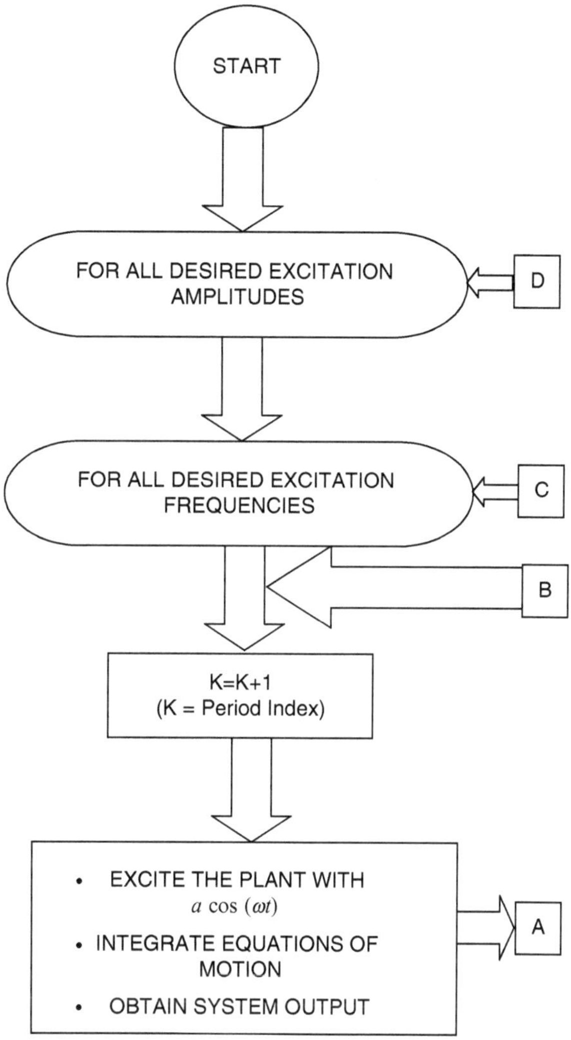

Fig. 2.3 Flow chart for describing function generator

$$T_{\mathrm{m}} = \begin{cases} T_{\mathrm{e}} - f_{\mathrm{v}}\dot{x}_1 - f_{\mathrm{c}}Sign\,(\dot{x}_1) & \text{if } |T_{\mathrm{e}}| > f_{\mathrm{c}} \\ T_{\mathrm{e}} - f_{\mathrm{v}}\dot{x}_1 - f_{\mathrm{c}}Sign\,(\dot{x}_1) & \text{if } \dot{x}_1 \neq 0 \\ 0.0 \text{ if } |T_{\mathrm{e}}| < f_{\mathrm{c}} & \text{and } \dot{x}_1 = 0 \end{cases} \tag{2.20}$$

$$T_{\mathrm{e}} = \begin{cases} m_1 V_{\mathrm{in}} & \text{if } |V_{\mathrm{in}}| \leq \delta \\ Sign(V_{\mathrm{in}}).\,(m_1\delta + m_2\,(|V_{\mathrm{in}}| - \delta)) & \text{if } |V_{\mathrm{in}}| > \delta, \end{cases} \tag{2.21}$$

where f_{c} Coulomb friction, 0.1Nm; f_{v} coefficient of viscous friction, 0.1Nm−s/rad; J moment of inertia, 0.01Kg − m^2; m_1 a slope, 5Nm /V; m_2 a slope, 1Nm /V; δ saturation break point, 0.5V.

Fig. 2.3 (continued)

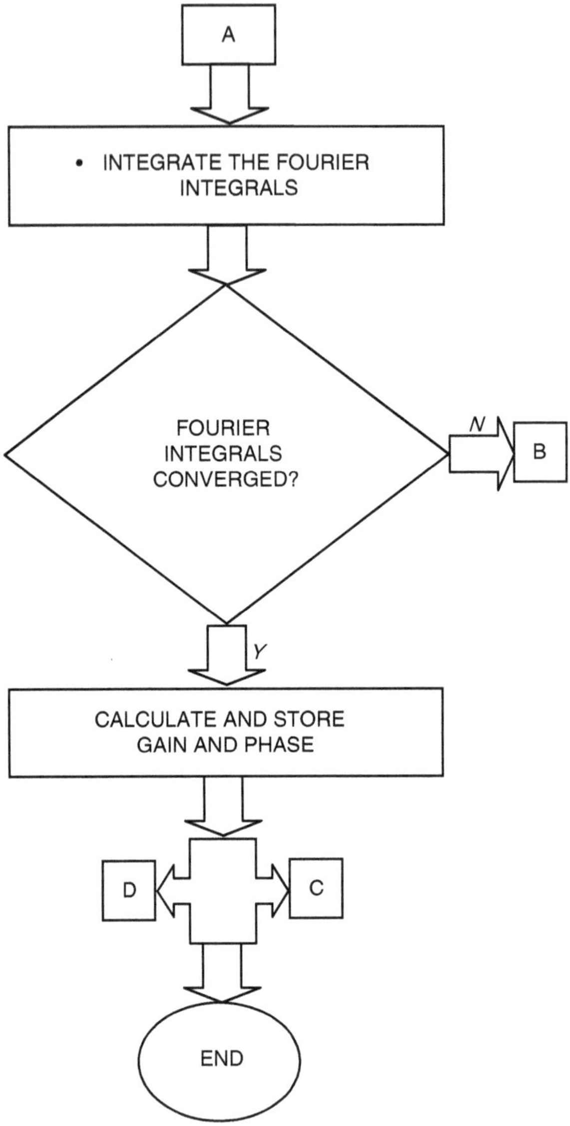

The computer model (Taylor and Strobel 1985) of this process in terms of a FORTRAN subroutine is given by Nassirharand 1987. By executing the MATLAB® function described in the previous section, the generated describing function models are obtained, and they are depicted in Figs. 2.5 and 2.6. The corresponding three-dimensional plots are given by Figs. 2.7 and 2.8.

From examination of Figs. 2.7 and 2.8, it is apparent that the gain and phase surfaces as a function of amplitude are not flat; hence, the system is not linear.

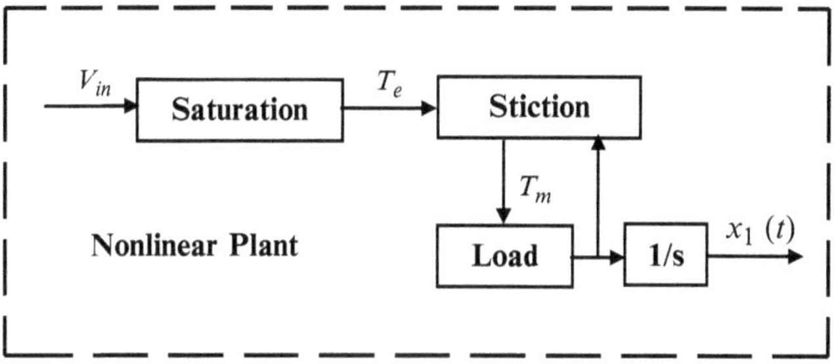

Fig. 2.4 The schematic model of the servomechanism example

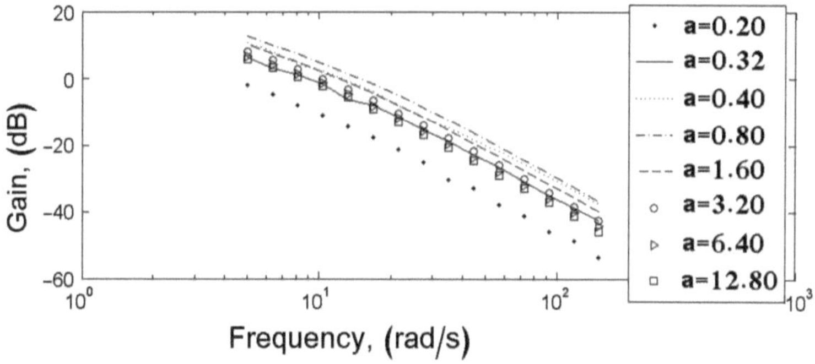

Fig. 2.5 The pseudo Bode gain plot (a is the amplitude of excitation)

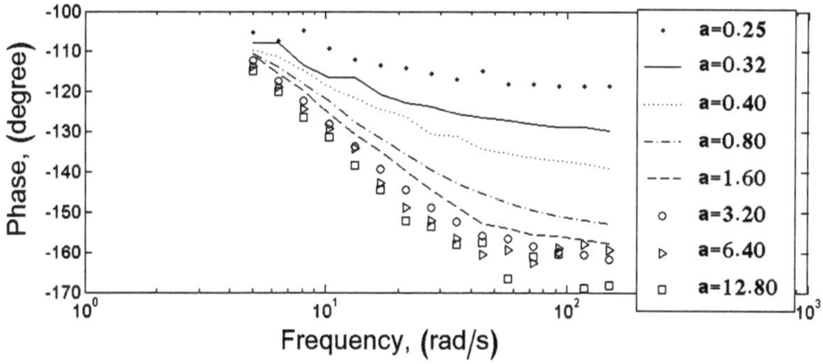

Fig. 2.6 The pseudo Bode phase plot (a is the amplitude of excitation)

2.2.2 *Multi-input Multi-output Case*

There are two important key points that must be kept in mind when characterizing the input/output behavior of a multivariable nonlinear process. The first one is that principle of superposition does not hold here; therefore, one is not able to determine the effects of each input on the corresponding outputs and then sum the effects.

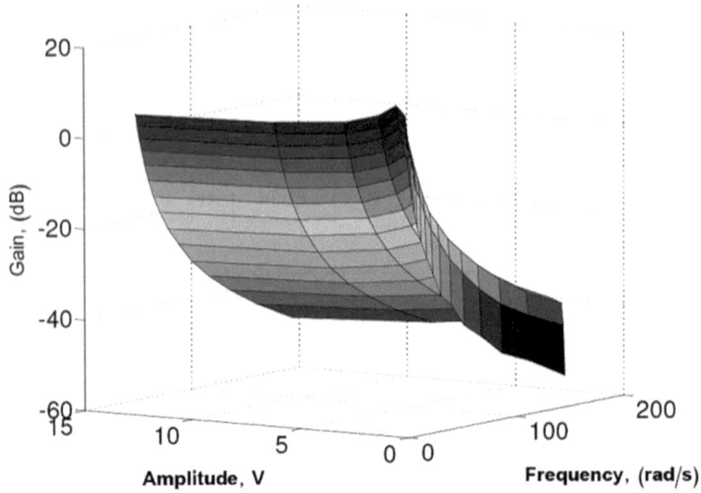

Fig. 2.7 The gain surface of the pseudo frequency response

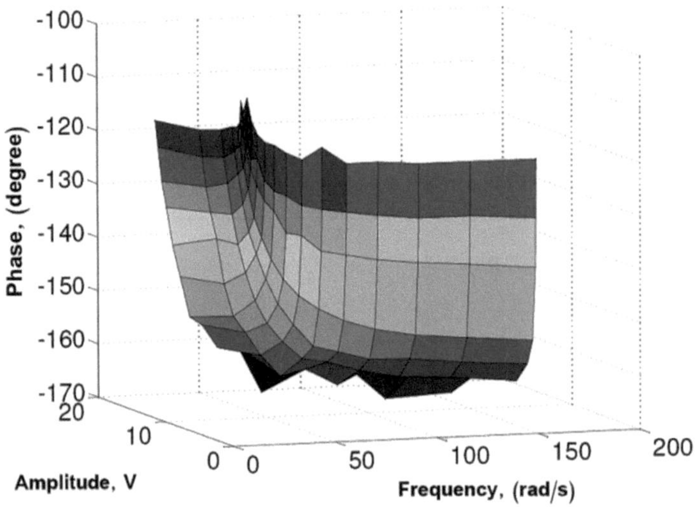

Fig. 2.8 The phase surface of the pseudo frequency response

Hence, all input channels must all be simultaneously activated. The second key point is that excitation frequencies of inputs could not be the same because the separate effects of each input on each output would be indeterminate.

The characterization approach proceeds in the following fashion. The nonlinear multivariable system given by Eqs. 1.1 and 1.2 is excited by input signals of the following general form:

$$u_p(t) = u_{0,p} + a_p \cos\left(\omega_p t + \theta_p\right), \quad p = 1, 2, ...m, \tag{2.22}$$

where u is the input, u_0 is the DC component of the input, a is the amplitude of excitation, ω is the frequency of excitation, θ is the phase shift, and p is the input channel index. Then, the dynamic equations of motion are numerically integrated to obtain the outputs as a function of time, $y_q(t), q = 1, 2, ..., l$. Then, the matrix Fourier integrals, denoted $I_{q,p}^{h,k}$, are evaluated.

$$I_{q,p}^{h,k} = \int_{(k-1)T}^{kT} y_q(t) \exp\left(-jh\left(\omega_p t + \theta_p\right)\right) \mathrm{d}t, \tag{2.23}$$

where h denotes the harmonic index, k denotes the period index of the Fourier integral, q is the output channel index, and T is the overall period of the input vector. Finally, the pseudo frequency response estimate, denoted by matrix $G_{q,p}^{1,k}$ is calculated as follows:

$$G_{q,p}^{1,k}\left(j\omega_q; u_0, a, \theta\right) = \frac{2}{a_p T} I_{q,p}^{1,k}. \tag{2.24}$$

In order to analyze the higher harmonic effects, one sets $h = 2, 3, ...$

The Fourier integrals are said to have converged if the following conditions are satisfied:

$$\left|\frac{M_{q,p}^{k} - M_{q,p}^{k-1}}{M_{q,p}^{k}}\right| < \varepsilon_{\mathrm{M}}, \tag{2.25}$$

$$\left|\varphi_{q,p}^{k} - \varphi_{q,p}^{k-1}\right| < \varepsilon_{\varphi}, \tag{2.26}$$

where $M_{q,p}^{k}$ and $\phi_{q,p}^{k}$ represent the log magnitude and phase of $G_{q,p}^{k}\,(j\omega; u_0, a, \theta)$.

2.2.2.1 Software

A MATLAB® function similar to the single-variable case is available whose primary inputs are (1) the SIMULINK® file name corresponding to the model of the multivariable nonlinear plant, (2) a matrix of excitation amplitudes, and (3) a matrix of excitation frequencies. The primary outputs are the matrices of real and imaginary parts of the describing function models as a function of excitation amplitudes and frequencies. At the nucleus of this software is a MATLAB® function that generates the matrix pseudo frequency response data for one amplitude and one frequency. This function is given in Fig. 2.9 (Nassirharand and Karimi 2002).

```
%
% FUNCTION getsidfm COMPUTES THE MATRIX PSEUDO
% FREQUENCY RESPONSE DATA FOR ONE SET OF INPUT
% AMPLITUDE AND ONE SET OF INPUT FREQUENCY ARRAYS.
% THE STATE-VARIABLE EQUATIONS ARE SOLVED
% USING ode45 FUNCTION, AND THE FUNCTION THAT
% IMPLEMENTS THE OUTPUT EQUATION MUST ALSO BE
% PROVIDED. THIS FUNCTION MAY BE MODIFIED TO ACCEPT
% THE TIME-DOMAIN SIMULATION DATA AS INPUT.
% FUNCTION INPUTS FOLLOW.
% sveq  a string variable containing the name of
%       the function that evaluates the rate of state-
%        Variables.
% w     frequency array of input sinusoids
% teta  phase shift array of the input sinusoids
% u0    DC value array of the input sinusoids
% ampl  amplitude array of input sinusoids
% kmax  maximum number of cycles to be considered
% nsv   number of state variables
%
% FUNCTION OUTPUTS FOLLOW.
%
% g      a two dimensional array whose first dimension
%        corresponds to the real part of pseudo
%        frequency response and the second dimension
%        corresponds to the imaginary part
% amer   a two dimensional array containing the error
%        in magnitude convergence condition (see
%        Equation (2.25)
% pher   a two dimensional array containing the error
%        in phase convergence condition (see Equation
%       (2.26)
% k      the index for the cycle that the Fourier
%        integrals have converged
%
function [g,amer,pher,k] =
                getsidfm(sveq,w,teta,u0,ampl,kmax,nsv)
  global amplp wp thetap u0p
  amplp=ampl;wp=w;thetap=teta;u0p=u0;
  lw=length(w);lampl=length(ampl);
% SPECIFY EPSILONS IN EQUATIONS (2.25) AND (2.26)
  amepsln=0.05;phsepsln=5;
```

Fig. 2.9 Listing for the multivariable sinusoidal input describing function generator. Source: (Nassirharand and Karimi 2002) © Elsevier

```
% COMPUTE PERIOD
T=getperiod(w);
% k IS THE INDEX FOR FOURIER INTEGRAL CYCLE
for k=1:kmax
   if(k==1)
         t0=0;
         for isv=1:nsv;x0(isv)=0;end
    else
         t0=tfinal;
         sx=size(x);isx=sx(1);
         for isv=1:nsv;x0(isv)=x(isx,isv);end
    end
    tfinal=k*T;tspan=[t0 tfinal];
% INTEGRATE STATE-VARIABLE EQUATIONS FOR ONE CYCLE
    [t,x]=ode45(sveq,tspan,x0);
% COMPUTE OUTPUT USING USER-SUPPLIED FUNCTION
% GETOUTPUT
    y=getoutput(x,t);
    sy=size(y);ly=sy(2);iconv=0;
    for q=1:ly;
        for p=1:lampl;
% COMPUTE THE REAL PART OF THE INTEGRAND OF FOURIER
% INTEGRAL
intgd1=y(:,q).*cos(w(p)*t+teta(p));
% INTEGRATE
fr=trapz(t,intgd1);
% COMPUTE THE IMAGINARY PART OF THE INTEGRAND OF
% FOURIER INTEGRAL
intgd2=-y(:,q).*sin(w(p)*t+teta(p));
%INTEGRATE
fi=trapz(t,intgd2);
%COMPUTE THE REAL PART OF THE PSEUDO FREQUENCY
%RESPONSE
            hr(q,p)=(2*fr)/(ampl(p)*T);
%COMPUTE THE IMAGINARY PART OF THE PSEUDO FREQUENCY
%RESPONSE
            hi(q,p)=(2*fi)/(ampl(p)*T);
```

Fig. 2.9 (continued)

```
% COMPUTE MAGNITUDE
am(q,p,k)=sqrt(hr(q,p)^2+hi(q,p)^2);
%COMPUTE PHASE
phi(q,p,k)=atan2(hi(q,p),hr(q,p));
 if(k>=2)
% COMPUTE EPSILON IN EQUATION (2.25)
 amchk(q,p)=abs( (am(q,p,k)-am(q,p,k-1)) )/
          (abs(am(q,p,k)));
% COMPUTE PHASE IN EQUATION (2.26)
 phchk(q,p)=abs(phi(q,p,k)-phi(q,p,k-1));
%CHECK CONVERGENCE
if(amchk(q,p) <= amepsln)
    if(phchk(q,p) <= phsepsln)
        iconv=iconv+1;
     end
 end
% END CALCULATION OF MATRIX EPSILONS
end
% END THE INPUT INDEX LOOP
End
% END THE OUTPUT INDEX LOOP
end
if(iconv==(ly*lampl))
% FOURIER INTEGRALS CONVERGED
        for q=1:ly;
           for p=1:lampl;
% STORE THE PSEUDO COMPLEX FREQUENCY RESPONSE
% VALUE IN g
              g(q,p)=complex(hr(q,p),hi(q,p));
% STORE ERROR IN MAGNITUDE AND PHASE
           amer(q,p)=amchk(q,p);pher(q,p)=phchk(q,p);
           end
        end
% EXIT FOURIER INTEGRAL COMPUTATIONS LOOP
        break
     end
```

Fig. 2.9 (continued)

```
   if(k==kmax)
% FOURIER INTEGRALS DID NOT CONVERGE
           for q=1:ly;
              for p=1:lampl;
% STORE THE LATEST VALUE OF MATRIX PASUDO FREQUENCY
% RESPONSE VALUE
                 g(q,p)=complex(hr(q,p),hi(q,p));
% STORE EPSILONS IN EQUATION (2.25) AND (2.26)
             amer(q,p)=amchk(q,p);pher(q,p)=phchk(q,p);
              end
           end
        end
%END THE FOURIER CYCLE LOOP
end

%
%FUNCTION getperiod COMPUTES PERIOD
%ASSUMES THE INPUT SINUSOIDS MEET AT SOME TIME
%
 function T=getperiod(w);
  ic=0;n=length(w);
  for j=1:n-1
     for i=j:n-1
        diff=abs(w(j)-w(i+1));
        ic=ic+1;
        TT(ic)=2*pi/diff
     end
  end
T=max(TT);
```

Fig. 2.9 (continued)

2.2.2.2 Example: An Automobile Engine

An idle speed control model used by Bengea et al. 2004 is utilized to demonstrate the idle speed controller design procedure and the typical results that may be achieved. This model is given below:

$$\frac{\mathrm{d}P_{\mathrm{m}}(t)}{\mathrm{d}t} + \frac{\eta_{\mathrm{v}} V_{\mathrm{d}}}{4\pi V_{\mathrm{m}}} N(t) P_{\mathrm{m}}(t) = K_1' \alpha(t), \tag{2.27}$$

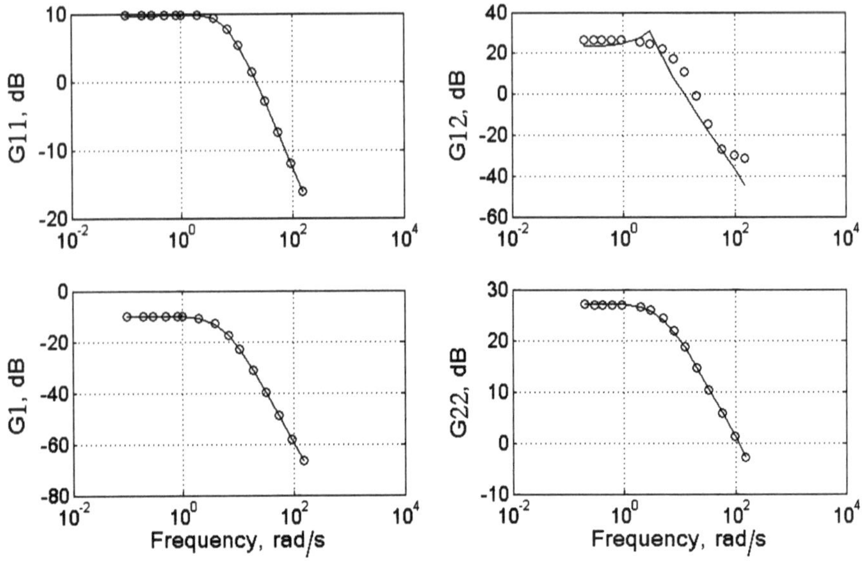

Fig. 2.10 Describing function model for the automobile engine – matrix gain plot. Source: (Teh and Nassirharand 2010a) © Sage

$$J\frac{\mathrm{d}N(t)}{\mathrm{d}t}+\xi N(t)=k_\tau P_\mathrm{m}(t-\tau)+k_\delta\delta(t)+\tau_\mathrm{f}(t), \tag{2.28}$$

where$J$ engine inertia ($0.0843\ \mathrm{N}-\mathrm{ms}^2/\mathrm{rad}$); k_δ a coefficient relating engine torque to the spark advance ($2\pm 0.05\ \mathrm{Nm}/\mathrm{deg}$); k_τ a coefficient relating engine torque to the intake manifold pressure ($0.57143\pm 0.015\ \mathrm{Nm}/\mathrm{kPa}$); K_1' a constant relating the idle throttle to the air mass flow rate past the throttle plate (110 ± 2.5 $(\mathrm{kPa}/\mathrm{s})/\mathrm{deg}$); $N(t)$ engine speed, dependence on t may be omitted for simplicity (rad/s); $P_\mathrm{m}(t)$ intake manifold pressure, dependence on t may be omitted for simplicity (Pa); t time variable (s); V_d engine displacement ($0.0046\ m^3$); V_m manifold volume ($0.0029\ \mathrm{m}^3$); $\alpha(t)$ idle bypass valve opening (deg); $\delta(t)$ spark timing advance (deg); η_v volumetric efficiency of the pumping process (0.55 ± 0.025 dimensionless); π 3.14 . . . .; τ induction to torque delay at nominal idle speed; N_0, (0.1 s); τ_f disturbance torque ($\mathrm{N}-\mathrm{m}$); ξ lumped damping coefficient ($0.592\pm 0.0125\ \mathrm{Nm}-\mathrm{s}$).

The parameter values and the associated uncertainties given in the table above correspond to those of the Ford V8 4.6 Liter engine. This model and the corresponding parameter values have been used quite frequently by many researchers (e.g., Bengea et al. 2004) as a benchmark problem to test their controller design procedures. Here, the same model parameters are used for illustration purposes.

A SIMULINK® model of Eqs. 2.27 and 2.28 is developed. Then, the output time histories for various sinusoidal inputs with $u_{0,1}=17.04\mathrm{deg}$, $u_{0,2}=0.2676\,\mathrm{deg}$, $a_1=0.01u_{01}\mathrm{rad}/\mathrm{s}$, $a_2=0.01u_{02}\mathrm{kPa}$, $\omega_1=[0.1, 0.2, 0.3, 0.5, 0.8, 1, 2, 4, 7, 11,$

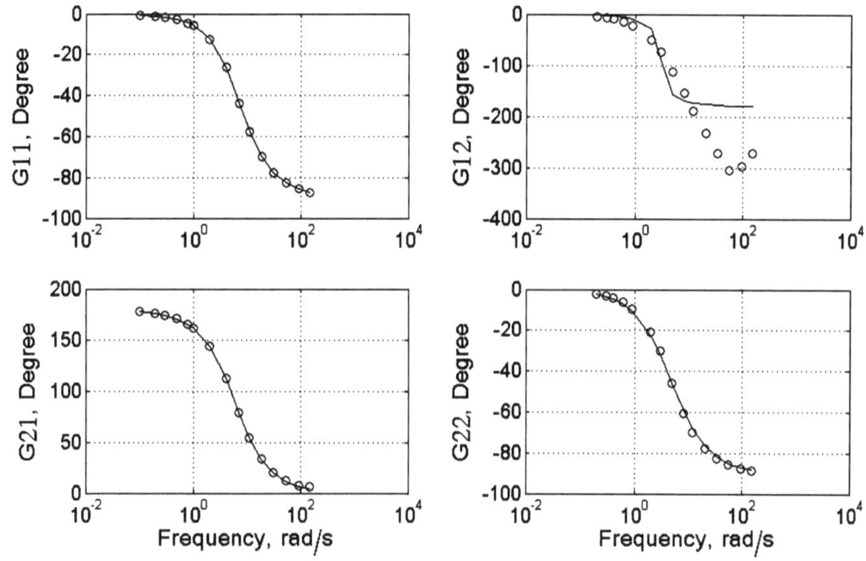

Fig. 2.11 Describing function model for the automobile engine – matrix phase plot. Source: (Teh and Nassirharand 2010a) © Sage

19, 32, 55, 94, 150]rad/s, $\omega_2 = [0.2, 0.3, 0.4, 0.6, 0.9, 2, 3, 5, 8, 12, 20, 33, 56, 95, 151]$rad/s, and $\theta_1 = \theta_2 = 0$ deg are generated. Generally speaking, the user should avoid frequencies that are simple multiples of one another (e.g., one is twice as much as the other); however, our experience with various problems in aerospace, Mechatronics, and automotive has revealed that in those specific cases no major flaws in the Fourier analysis were present when isolated simple multiples of frequencies were used. In continuation, the Fourier integrals are evaluated when the outputs reach steady state; then, the SIDF model of the I.C. engine is generated. This quasi-linear model along with system identification results is depicted in Figs. 2.10 and 2.11 (Teh and Nassirharand 2010a). In these figures, the small circles correspond to SIDF models, and the solid lines correspond to the linear fits obtained using *invfreqs* function of MATLAB®.

Chapter 3
Single-Range Controller Design

3.1 Procedure

In this part, a systematic procedure for design of single-range controllers is presented; this algorithm is a modified version of its original version outlined by Taylor (1983). The nonlinear and dynamic mathematical model of the system, which may include both soft (continuous) and hard (discontinuous) nonlinearities, is assumed to be available. The systematic controller design procedure is based on describing function models of the plant coupled with any linear controller design technique. It should be emphasized that, in general, small-signal models for this class of nonlinear systems do not exist.

The control system design procedure is systematic, and it involves six steps as given by the following algorithm.

Step 1 relates to user-defined performance measures, and the user must identify the model of the process that he wants to mimic. In most cases, the desired linear model would be a second-order transfer function that possesses the desired dominant poles, and satisfy the desired steady state error conditions. The user may translate the time and/or frequency performance measures to the desired natural frequency and the desired damping ratio by considering the performance specification equations noted in Rowland (1986); then the steady state error specifications determine if a zero is also required. The zero may be determined from relations for definition of a steady state error.

Step 2 requires the knowledge of the operating regimes of interest. Note that unlike operating points, operating regimes are characterized by the range of expected amplitudes and frequencies of the excitation signal. For example, for the case of regulating the pressure of the combustion chamber of a liquid propellant engine, one may use the steady-state values of the positions of the regulator valve piston such that those values result in the desired values of the combustion chamber pressures (Nassirharand and Karimi 2004a). Note that the mentioned steady-state values of the position of the regulator valve piston are known by design; alternatively, the control system designer may determine the desired steady

A. Nassirharand, *Computer-aided Nonlinear Control System Design: Using Describing Function Models*, DOI 10.1007/978-1-4471-2149-7_3,
© Springer-Verlag London Limited 2012

state values of the position of the regulator valve control piston by simulation and by noting those values of the position of the regulator valve control piston that result in the desired values of the combustion chamber pressure. The expected frequencies of excitation are usually determined from the knowledge of the natural frequency of the system. As a rule of thumb, the lower range would be about two decades below the natural frequency of the system, and the upper frequency range would be about one decade above the natural frequency of the system. Alternatively, the user may initially set the lower and upper frequency ranges to 0.1 rad/s and 50 rad/s, respectively. Then, the pseudofrequency response plots are generated. If the low-frequency response portion is not characterized, then the lower frequency range may be lowered by one decade and so on until the low-frequency response portion is characterized. The upper frequency range is also increased if the high-frequency gain of the pseudofrequency response is substantial. Then the Fourier-based approach described in the previous chapter would be used to obtain the describing function models.

Algorithm 3.1

1. Identify the desired reference linear model.
2. Obtain the describing function models of the plant, and select a nominal model.
3. Identify a linear model of the nominal frequency domain model of the previous step (optional).
4. Design a controller based on any available linear control theory.
5. Verify design, and use optimization to tune the controller gains as necessary.
6. If the results are satisfactory, then stop; otherwise consider designing a dual-range linear controller.

In step 3, once the describing function models are obtained, one of these models is selected as the nominal model. Normally, the nominal model would be the case in which the excitation signal would correspond to the normal operating conditions of the plant in the absence of disturbances, in the absence of loss of sensory devices, or in the absence of any perturbations of the plant. Note that assumption is that one describing function model of the system adequately describes the dynamic behavior of the system; otherwise, a dual-range or a multi-range controller design approach may have to be employed (see Chaps. 4 and 5). In any case, it is recommended to study the effectiveness of a single-range controller before a more complicated controller design procedure such as a dual-range or a multi-range nonlinear control approach is adopted. In applications that nonlinearity effects are not dominant, the spread in describing function models would not be significant, and therefore, a single-range linear controller would suffice.

Step 4 requires identification of a linear model whose frequency response data matches that of the nominal describing function model of the previous step. Since the describing function models are representation of nonlinear systems, the standard relation between the two components of the frequency response data that exists for linear systems does not hold for describing function models. Therefore, care must be taken when fitting the pseudofrequency response data. The outcome of this step is a linear model described in terms of a transfer function.

Step 5 is to use the linear model of the previous step, and to interface with any one of the available linear controller design techniques to design a linear controller. For example, one may use factorization theory coupled with analytical optimization, and obtain a controller design equation of the following form (see Appendix B for details):

$$A'X = B \tag{3.1}$$

where A' and B are known, and they are in terms of coefficients of coprime factors of the linear model of the previous step, the coefficients of coprime factors that are solution to the Bezout identity, the coefficients of the desired linear model of step 1, and frequency range of interest; vector X contains the coefficients of the function parameter, $r(s)$, that must be substituted in the celebrated Youla parameterization equation (see Eq. B.5) to obtain the desired controller (Youla et al. 1976).

Finally, in step 5, the design is verified; in some cases, the controller parameters may have to be tuned. In such cases, one may use optimization for this purpose.

3.2 Example – Single-Variable Case: A Decoupled Liquid Propellant Engine

A case study of the sort encountered in combustion chamber pressure control of a liquid propellant engine (LPE) is considered. A simplified schematic diagram of the LPE is shown in Fig. 3.1. The operation of the engine starts by an electric command that starts combustion in the starter. This combustion generates pressurized gas that opens the diaphragm valves (not shown) below the fuel and oxidizer tanks, and it also turns the turbine. Consequently, the shaft of the turbopump starts rotating, and the fuel and oxidizer pumps start pumping fuel and oxidizer into the system. Now, consider the two fuel and oxidizer paths. In the fuel path, fuel enters the stabilizer control valve, and the output of this valve is a controlled amount of fuel toward the combustion chamber and the gas generator. At the same time, in the oxidizer path, oxidizer passes through the shut-down valve (not shown), and it enters the combustion chamber. This fuel and oxidizer that have entered the combustion chamber produce the required thrust via combustion. This thrust must be controlled. The amount of oxidizer that enters the gas generator indirectly controls the thrust, and the desired amount of oxidizer that must flow through the gas generator is utilized via the regulator control valve. This oxidizer along with the fuel that have entered the gas generator from the fuel path produce a combustion; this combustion results in a pressurized gas that keeps the turbine running when the starter leaves

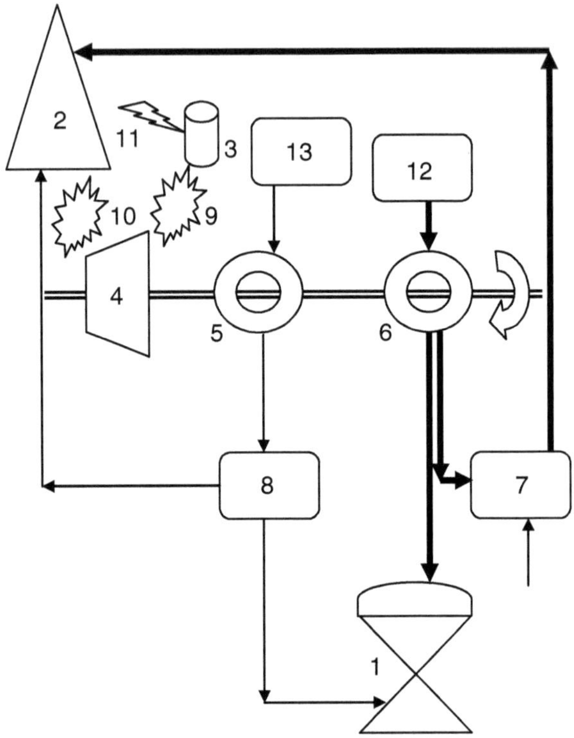

Fig. 3.1 A simplified schematic diagram of the liquid engine (Key is given in Table 3.1)

Table 3.1 Key for Fig. 3.1

No.	Description
1	Combustion camber
2	Gas generator
3	Starter
4	Turbine
5	Fuel pump
6	Oxidizer pump
7	Regulator valve
8	Stabilizer valve
9	Starter combustion products
10	Gas generator combustion products
11	Electric command
12	Oxidizer tank
13	Fuel tank

the circuit. Depending on the value of the angular speed of the turbine, controlled amounts of fuel and oxidizer flow toward the combustion chamber.

The combustion chamber pressure control system is one of the most important parts of a liquid propellant engine. A class of regulators, which is used to control the combustion chamber pressure, is composed of complicated hydraulic control valves. These control valves have many disadvantages (Nassirharand and Karimi 2004a). For example, development of nonlinear and dynamic mathematical model of such control valves is very time consuming and difficult. When such models are developed and incorporated with the rest of the engine model, the execution time of the simulation code is considerably increased. Another disadvantage is that the control valve may accept only one set point (the set point is the desired pressure value of the combustion chamber that the control system must maintain).

Therefore, it is desirable to design an alternative pressure control system which would involve microprocessor-based regulators. Such pressure regulators could accept various set points by software. The advantages of regulators that could accept various set points are (1) the final desired pressure value could be reached at specified stages in order to reduce the effects of water-hammer as well as excessive overshoot of the pressure inside the combustion chamber and (2) to reduce the generated impulse after the engine shut down command is issued; again, in this case, the combustion chamber pressure is reduced from its steady-state value to zero at specified stages.

The design of the regulator control loop is fully based on dynamic and nonlinear mathematical model of the engine (Karimi et al. 2003). In order to develop the mathematical model of the engine, four groups of elements are identified. Those element groups are (1) valve, (2) turbopump, (3) combustion chamber and gas generator, and (4) connections (e.g., pipes). In this treatment, the mathematical model of the engine is put into perspective by selecting one element from each group, and the mathematical model of that element is developed.

Without loss of generality, an orifice model could be used for valves and connection pipes. The fuel pump model follows, and the oxidizer pump equations are similar:

$$\dot{m}_5 = \rho_{fu} Q_5 \tag{3.2}$$

$$TQ_5 = \frac{\gamma_5 H_5 Q_5}{\omega_5 \eta_5} \tag{3.3}$$

$$H_5 = \left(a_5 + b_5 Q_5 - c_5 Q_5^2\right)/g \tag{3.4}$$

$$a_5 = A_5 \omega_5^2 \tag{3.5}$$

$$A_5 = \left(0.97 + 0.8 q_{5,p}\right)\left(\frac{H_{5,p}}{\omega_{5,p}^2}\right) \tag{3.6}$$

$$q_{5,p} = \frac{1}{1 + \frac{2\pi H_{5,p} b_{5,2} \tan(\beta_{5,2\pi})}{Q_{5,p}\omega_{5,p} K_{5,z}\eta_{5,zp}}} \tag{3.7}$$

$$b_5 = B_5 \omega_5 \tag{3.8}$$

$$B_5 = \left(0.325 - 0.8q_{5p}\right)\left(\frac{H_{5,p}}{\omega_{5,p}^2}\right)\left(\frac{Q_{5,p}}{\omega_{5,p}}\right)^{-1} \tag{3.9}$$

$$c_5 = C_5 \tag{3.10}$$

$$\mathrm{C}_5 = 0.296\frac{\left(H_{5,p}/\omega_{5,p}^2\right)}{\left(Q_{5,p}/\omega_{5,p}\right)^2}, \tag{3.11}$$

where

A_5 = an intermediate variable for fuel pump head equation
a_5 = coefficient of the fuel pump head equation
B_5 = an intermediate variable for fuel pump head equation
b_5 = a coefficient of the fuel pump head equation
$b_{5,2}$ = the exit width of fuel pump impeller
C_5 = an intermediate variable for fuel pump head equation
c_5 = a coefficient of the fuel pump head equation
H_5 = the fuel pump head
$H_{5,p}$ = the nominal fuel pump head
$K_{5,z}$ = the effect coefficient of fuel pump blade
$\dot{m}_5$ = the mass flow rate of fuel pump
Q_5 = the volumetric flow rate of fuel pump
$Q_{5,p}$ = the nominal volumetric flow rate of fuel pump
$q_{5,p}$ = an intermediate parameter in terms of nominal values of fuel pump
TQ_5 = the moment of fuel pump
$\beta_{5,2\pi}$ = exit angle of fuel pump blade
$\gamma_5 = \rho_{fu} \cdot g$
η_5 = the efficiency of fuel pump
$\eta_{5,p}$ = the nominal efficiency of fuel pump
$\eta_{5,zp}$ = the hydraulic efficiency of fuel pump
ω_5 = the turbopump angular speed
$\omega_{5,p}$ = the nominal angular speed of turbopump
ρ_{fu} = the fuel density.

A simple combustion model is used as below:

$$\alpha\frac{\mathrm{d}P_1}{\mathrm{d}t} + \beta P_1 = \dot{m}_{1,fu}\left(t - \tau_1\right) + \dot{m}_{1,ox}\left(t - \tau_1\right), \tag{3.12}$$

where

$\dot{m}_{1,ox}$ = the oxidizer mass flow rate
P_1 = the combustion chamber pressure
α, β = functions of pressure and mixture ratio; for a given P_1 and mixture ratio, the values of α and β are computed off-line from energy equations, and

they are incorporated as tables of values into the computer model. Then, as simulation time marches ahead, the needed mentioned values for a given P_1 and mixture ratio are obtained by linear interpolation.

τ_1 = the combustion delay time.

In this part, a describing function approach coupled with factorization theory is used for design of the combustion chamber pressure control system. The primary reason for this selection is that engine model is of the form given by state variable equations of the form given by Eqs. 1.1 and 1.2, and describing function approach is inherently capable of handling such nonlinear models.

The problem statement for the pressure controller of the combustion chamber can be stated as follows – given the computer model of a liquid propellant engine, how does one design a controller that would control the pressure of the combustion chamber?

Algorithm 3.1 is applied to this problem. The nonlinear and dynamic computer model of the liquid propellant engine is utilized to design a controller for the combustion chamber. The controller design procedure steps are executed as follows. In step 1, the desired transfer function is identified in terms of a linear second-order transfer function that exhibits the desired behavior. The identified transfer function is of the following form:

$$h^D_{y,u} = \frac{100}{s^2 + 36s + 100}. \tag{3.13}$$

In step 2, the describing function models are obtained as outlined in Chap. 2, and these models are depicted in Fig. 3.2 (Nassirharand and Karimi 2004a). Notice that spread in these models is not significant, and therefore, it may be concluded that a single-range linear controller might suffice. The describing function model that is close to the nominal operating conditions was selected for system identification purposes of the next step. In step 3, the MATLAB® function *invfreqs* is used to identify a linear model. The identified linear model is of the following form:

$$G(s) = \frac{138.52}{s^2 + 52.81s + 253.47}. \tag{3.14}$$

The quality of this fit is shown in Fig. 3.3 (Nassirharand and Karimi 2004a). In step 4, Eq. B.26 is solved with

$$m = n = 3, \omega_1 = 0.01, \omega_2 = 5.$$

The following controller is designed:

$$C(s) = \frac{0.7219s^2 + 38.12s + 183.00}{s^2 + 36s}. \tag{3.15}$$

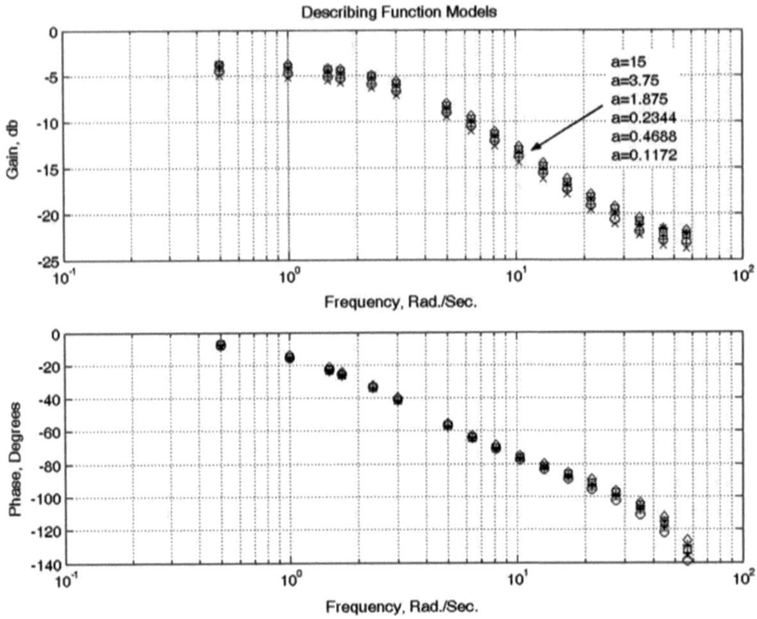

Fig. 3.2 SIDF models of the liquid propellant engine. Source: (Nassirharand and Karimi 2004a) © Sharif University of Technology

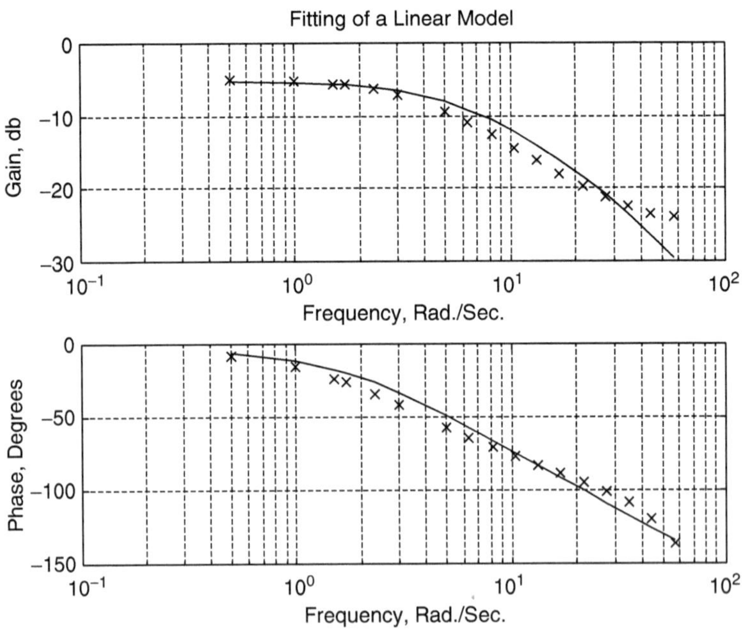

Fig. 3.3 Linear system identification. Source: (Nassirharand and Karimi 2004a) © Sharif University of Technology

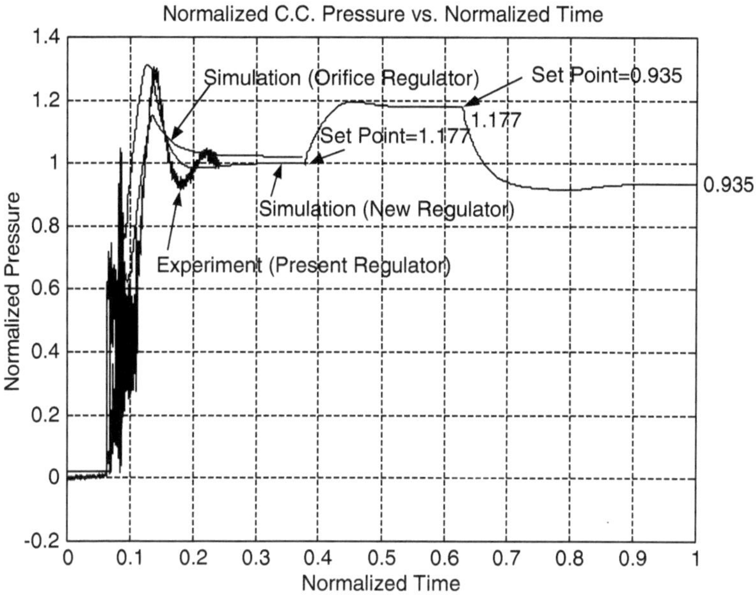

Fig. 3.4 Design verification. Source: (Nassirharand and Karimi 2004a) © Sharif University of Technology

The verification results are shown in Fig. 3.4 (Nassirharand and Karimi 2004a). In the beginning, the pressure set point is set to one unit, and later this set point is arbitrarily changed to 1.177 and 0.935 units in order to verify that regulator could accept various set points. In Fig. 3.4, there are three curves: (1) the experimental results of the engine with the present hydraulic regulator, (2) the simulation results of the engine with the present regulator modeled as an orifice, and (3) the simulation results of the engine with the new regulator. By comparison of these three curves, the following important point may be noted. One of the problems associated with the present pressure control system is that response overshoots. The curve corresponding to the new regulator design with second set point demonstrates that initial overshoot is not due to the regulator. In fact, the initial overshoot is due to the design of the starter of the engine. With the new design, one may redesign the starter to bring the combustion chamber pressure to a lower value, and then the new regulator would be able to bring the pressure value to its desired value without any significant amount of overshoot as is demonstrated in Fig. 3.4 (see the behavior corresponding to the second set point command of the curve entitled "Simulation + New Regulator"). The robustness test is performed by causing a 20% reduction in pumps' heads; the results are shown in Fig. 3.5 (Nassirharand and Karimi 2004a). From this figure it is concluded that design is also robust because describing function models are one of the most effective models to achieve a robust design.

For comparison purposes, in step 4 a H_∞ controller is designed (Nassirharand and Mousavi Firdeh 2009a). The structure of the H_∞ control feedback system is

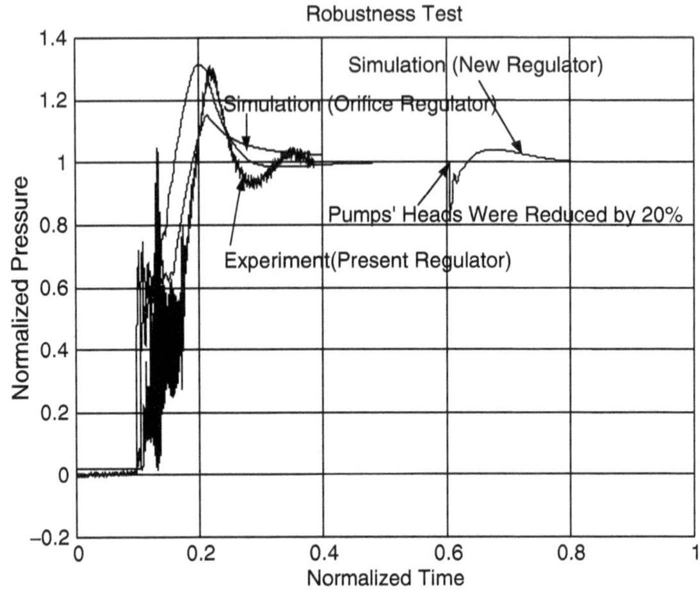

Fig. 3.5 Response with pumps' failures. Source: (Nassirharand and Karimi 2004a) © Sharif University of Technology

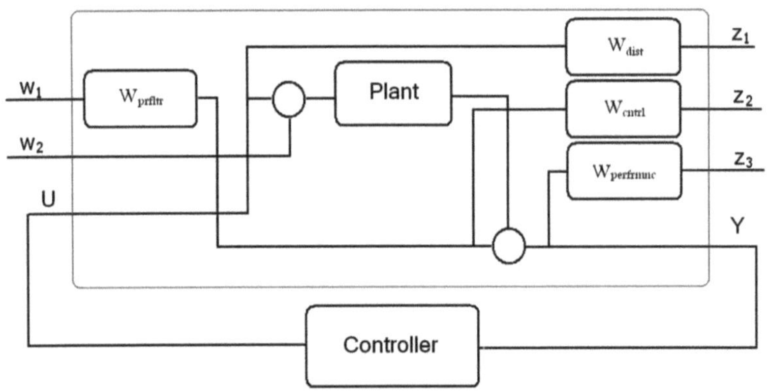

Fig. 3.6 Structure of the H_∞ feedback system. Source: (Nassirharand and Mousavi Firdeh 2009a) © Elsevier

shown in Fig. 3.6. In this figure, Ws are the weights, Z_1 is the weighted disturbance output signal, Z_2 is the weighted control signal, Z_3 is the weighted error signal, U is the control signal, and Y is the error signal. The following remarks are in order:

1. In order to achieve a bandwidth of 4 rad/s, a prefilter, $W_{\text{prfltr}} = \frac{4(s+250)}{250(s+4)}$, is specified; the zero of the prefilter is specified to satisfy the required H_∞ constraints (Doyle et al. 1989).

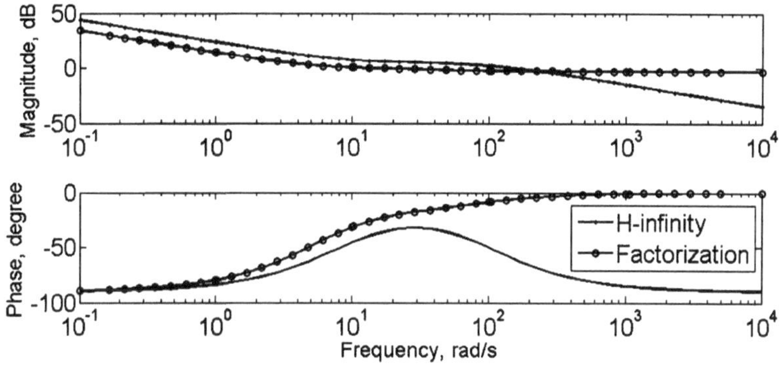

Fig. 3.7 Controller frequency response comparison. Source: (Nassirharand and Mousavi Firdeh 2009a) © Elsevier

2. It is desired to have a steady-state error of less than 1%; therefore, a performance weighting function, $W_{\text{perfrmnc}} = \frac{0.1(s+4)}{(s+0.01)}$, acting on the error signal is selected.
3. Considering a multiplicative uncertainty, a disturbance weighting function, $W_{\text{dist}} = \frac{2(s+9)}{(s+90)}$, for robust stability is specified. This reflects a 20% uncertainty for frequencies under 9 rad/s and a 200% uncertainty for frequencies above 90 rad/s.
4. In order to compensate for erosion and high levels of friction in mechanical parts, a control signal weighting function, $W_{\text{cntrl}} = \frac{3.5(s+12)}{(s+95)}$, is considered. This weighting function reflects the fact that is desired to weaken the control signals above approximately 15 Hz by a factor of 30%; it also reflects that it is desired to consider a weak penalty for low-frequency control signals.

A reduced order model of the H_∞ controller that is designed using the *hinfsys* function of MATLAB® software follows:

$$C(s) = \frac{182.3s + 1521}{s^2 + 99.4s}. \tag{3.16}$$

The performance of the H_∞ controller and that of the controller obtained using a factorization approach with the nonlinear plant is compared in Figs. 3.7, 3.8, 3.9, and 3.10 (Nassirharand and Mousavi Firdeh 2009a). From these figures, the following conclusions may be drawn:

1. The frequency response of the H_∞ controller (see Fig. 3.7) is more desirable because it attenuates the high-frequency signals; therefore, it is capable of being fairly insensitive to high-frequency noise in the system as demonstrated by Fig. 3.8. It is apparent that control signal of the H_∞ controller is less noisy than that of the factorization controller.
2. With reference to Fig. 3.9, the step response of the H_∞ controller is more desirable because it does not overshoot. Pressure overshoot could be harmful for the operation of the combustion chamber. Of course, the overshoot in the

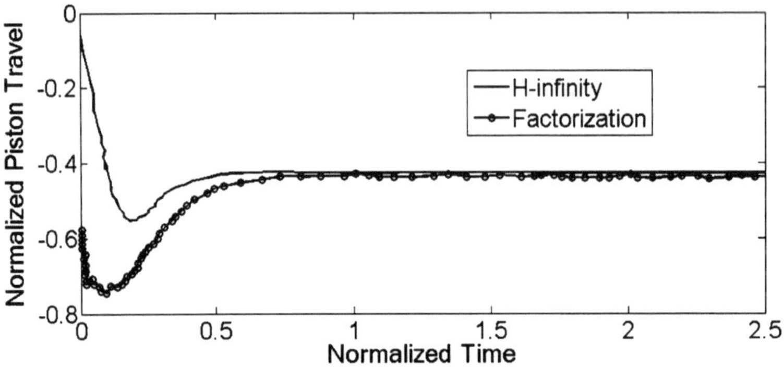

Fig. 3.8 Control signal comparison at the presence of measurement noise. Source: (Nassirharand and Mousavi Firdeh 2009a) © Elsevier

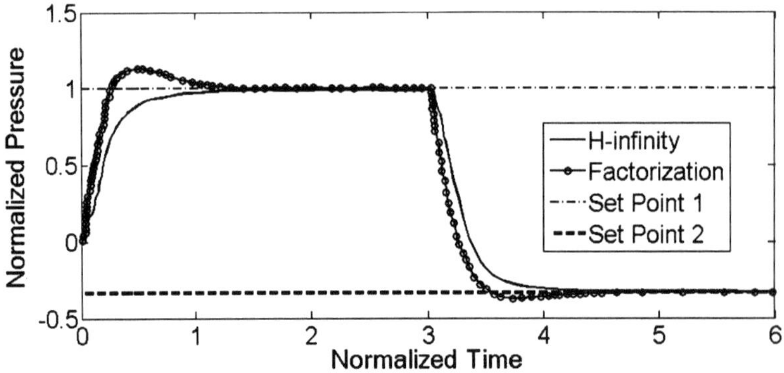

Fig. 3.9 Normalized step response comparison. Source: (Nassirharand and Mousavi Firdeh 2009a) © Elsevier

factorization design could be removed at the expense of an increase in the corresponding settling time.

3. With reference to Fig. 3.10, the control signal of the H_∞ controller is more desirable because it does not have the initial abrupt decrease in value.

3.3 Example – Multivariable Case: Idle Speed Control Problem

Algorithm 3.1 is applied to the idle speed control of uncertain automobile engines with time delay (Teh and Nassirharand 2010a). It is shown that results compete with the observer-based design of Bengea et al. (2004). Stability is demonstrated by successful generation of describing function models of the final nonlinear closed-loop system.

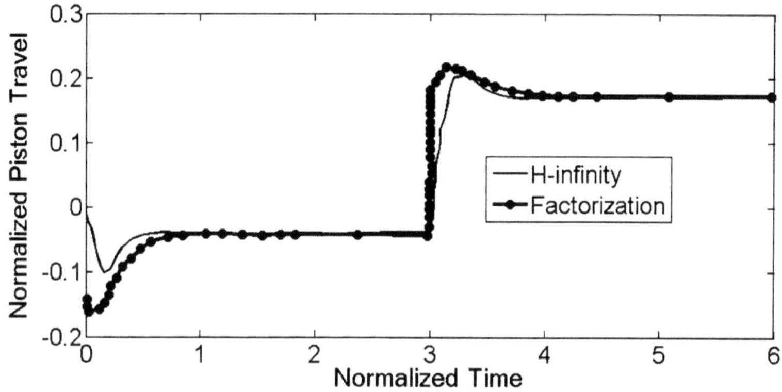

Fig. 3.10 Control signal comparison without measurement noise. Source: (Nassirharand and Mousavi Firdeh 2009a) © Elsevier

This problem may be classified as nonlinear, multivariable, uncertain, and distributed at the presence of load torque disturbances. Many researchers have developed various solution methods (see references in Teh and Nassirharand 2010a). These methods consider the following approaches:

(a) Linearizing the model and applying linear control theory
(b) Ignoring the time delay
(c) Approximating the time delay
(d) Utilizing an estimator or an observer
(e) Applying robust control techniques
(f) Applying neural networks
(g) Applying fuzzy control method
(h) Applying sliding mode or variable structure methods
(i) Applying optimal control technique
(j) Applying adaptive control theory

Except the standard methods mentioned in (a)–(c) above, all mentioned methods are relatively complicated. Those methods also require a high level of expertise in the utilized control method. These design methods usually involve invoking heuristics, and they require subjective judgment of the designer. The problems associated with the standard technique of linearizing the plant are well known. For example, the final control system would be highly sensitive to the operating point of interest (i.e., for standard plant perturbations, the closed-loop system may become unstable).

The use of mentioned complicated controller design methods is usually justified if the classical control system design methods do not produce adequate level of robustness, stability, and/or performance (Nassirharand et al. 1988). In continuation, a comprehensive explanation about the time delay of the IC engine during the design procedure is given. The presence of the time delay turns the plant into a distributed system. Such a system description would fit the class of nonlinear

systems considered herein as demonstrated by Eqs. 1.1 and 1.2. The effects of time delay and distributive nature of the plant are captured in the output time domain histories mentioned in step 2. Then these output time domain histories are transformed to their frequency domain counterparts by evaluation of the Fourier integrals as described earlier. Hence, the generated frequency domain model includes the effects of time delay. This frequency domain model is the basis of controller design in step 4. Finally, approximations made in obtaining the frequency domain model (including those related to the time delay approximations) are compensated for by tuning the gains of the elements of the matrix controller as described in step 5. Finally, the performance of the final synthesized controller with the actual nonlinear plant with the time delay is verified.
Algorithm 3.1 is executed as follows:

Step 1 – With reference to the work by Bengea et al. (2004), and the multivariable design approach outlined in Appendix B, the desired transfer function is selected for both main control loops:

$$h_{\mathrm{d}}(s) = \frac{225}{s^2 + 27s + 225}. \tag{3.17*}$$

Step 2 – The SIDF models are generated as outlined in Chap. 2, and these results along with linear system identification task of the following steps are shown in Figs. 2.10 and 2.11.
Step 3 – MATLAB® is used, and the following matrix transfer function is identified:

$$\mathbf{G}(s) = \begin{bmatrix} g_{11}(s) & g_{12}(s) \\ g_{21}(s) & g_{22}(s) \end{bmatrix}, \tag{3.18*}$$

where

$$g_{11}(s) = \frac{23.6799s + 59.1867}{s^2 + 9.4742s + 19.3981} \tag{3.19*}$$

$$g_{12}(s) = \frac{137.0035}{s^2 + 1.3259s + 9.0378} \tag{3.20*}$$

$$g_{21}(s) = \frac{-10.8771}{s^2 + 11.2327s + 34.1192} \tag{3.21*}$$

$$g_{22}(s) = \frac{111.4211}{s + 4.8593}. \tag{3.22*}$$

*Source: (Teh and Nassirharand 2010a) © Sage

Step 4 – From the application of linear lead-lag controller design approach outlined in Appendix A (with $\omega_1 = 0.01$ and $\omega_2 = 10$), the following compensator is designed:

$$c_{11}(s) = \frac{2.654s^2 + 1744s + 11930}{190.4s^2 + 4828s}. \tag{3.23*}$$

Similarly, with $h_d(s)$ as before, another compensator denoted $c_{22}(s)$ is designed:

$$c_{22}(s) = \frac{13.64s + 59.886}{6.52s^2 + 176.1s}. \tag{3.24*}$$

Now, use Eq. B.2 and obtain the off diagonal terms of the 2 by 2 matrix compensator transfer function, $C(s)$, as follows:

$$c_{12}(s) = \frac{-18.69s^3 - 259.1s^2 - 1140s - 1592}{1.544s^5 + 47.6s^4 + 178.6s^3 + 549.9s^2 + 941.9s} \tag{3.25*}$$

$$c_{21}(s) = \frac{2.789s^3 + 1911s^2 + 22200s + 63070}{2121s^4 + 77620s^3 + 676600s^2 + 1835000s}. \tag{3.26*}$$

Step 5 – Design verification reveals that controller gains must be tuned. To tune the gains, the following objective function is used:

$$\begin{aligned} E &= [N_1(t) - N_d(t)]^2 + [N_2(t)]^2 \\ &\quad + [P_{m1}(t) - P_{md}(t)]^2 + [P_{m2}(t)]^2. \end{aligned} \tag{3.27*}$$

The idea behind the selection and applicability of the above optimization criterion is explained next. The first and the third terms of the objective function assure tracking of the command signals of the two channels (speed and manifold pressure); the second and fourth terms of the objective function assure decoupling of the speed and manifold pressure axis. This objective function is also used in engine propulsion control as demonstrated in Nassirharand and Karimi (2004b).

For this example, $\mathbf{C}(s)$ is of the following form:

$$\mathbf{C}(s) = \begin{bmatrix} K_1 c_{11}(s) & K_2 c_{12}(s) \\ K_3 c_{21}(s) & K_4 c_{22}(s) \end{bmatrix}. \tag{3.28*}$$

In this research, the simplex optimization method implemented by the *fminsearch* function of MATLAB® is utilized.

---

*Source: (Teh and Nassirharand 2010a) © Sage

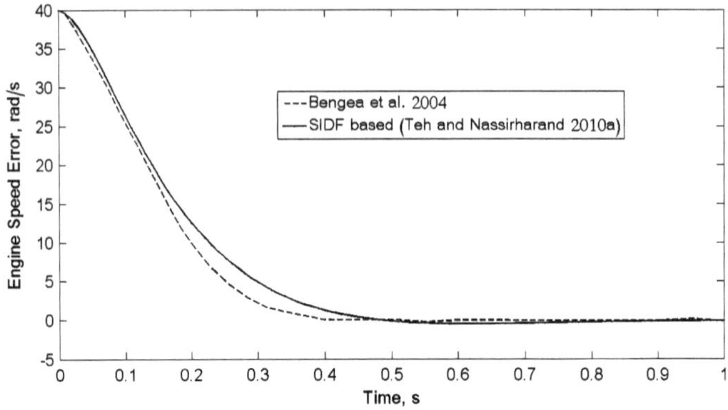

Fig. 3.11 Design comparison – speed axis. Source: (Teh and Nassirharand 2010a) © Sage

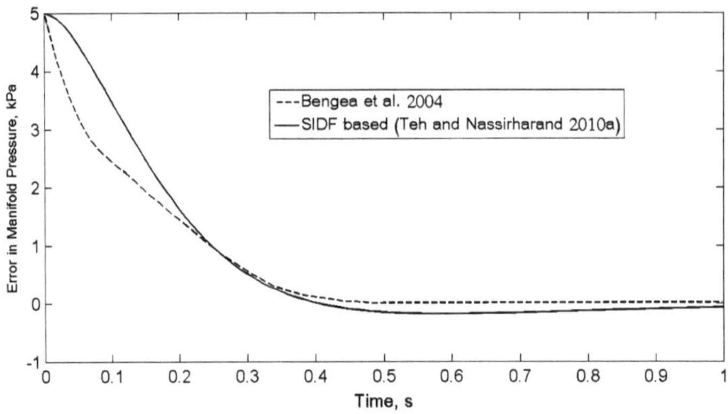

Fig. 3.12 Design comparison – pressure axis. Source: (Teh and Nassirharand 2010a) © Sage

The gains of the controller are *tuned* by minimizing objective function (3.27) with $N_2(t) = P_{m2}(t) = 0$. The optimum values of these gains follow:

$$K_1 = 0.6933, K_2 = 0.1915, K_3 = 0.0015, K_4 = 0.5172.$$

Finally, the design is verified by a digital simulation of the linear controller and the nonlinear plant; the results are compared with those obtained by the method of Bengea et al. (2004), as shown in Figs. 3.11, 3.12, 3.13, and 3.14. It is apparent that the classical design presented herein competes with the more complicated observer-based approach. Furthermore, the design is also verified to be robust with respect to engine parameter uncertainties and torque load disturbances. This is done by simulating the linear controller and the nonlinear plant with a disturbance load torque of 10 $N - m$ as well as nominal, upper limits, and lower limits of the engine parameters. These results are shown in Figs. 3.15 and 3.16.

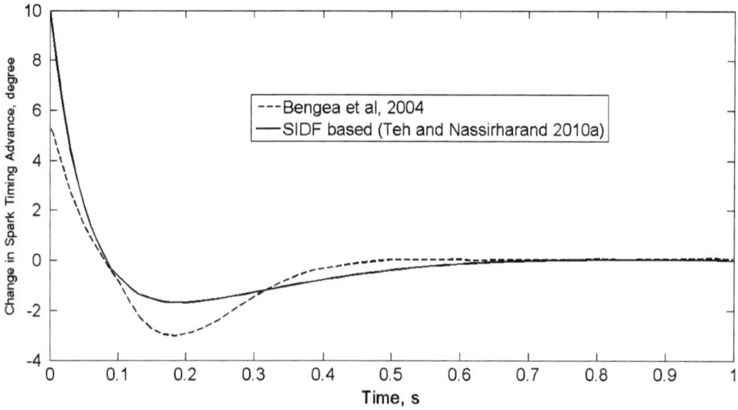

Fig. 3.13 Design comparison of control signals – spark timing advance. Source: (Teh and Nassirharand 2010a) © Sage

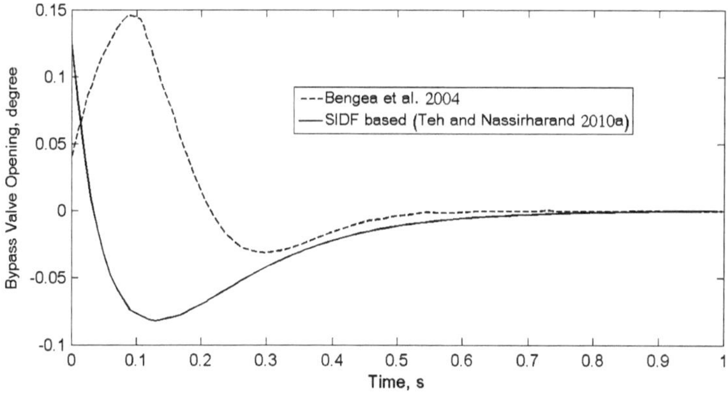

Fig. 3.14 Design comparison of control signals – idle bypass valve opening. Source: (Teh and Nassirharand 2010a) © Sage

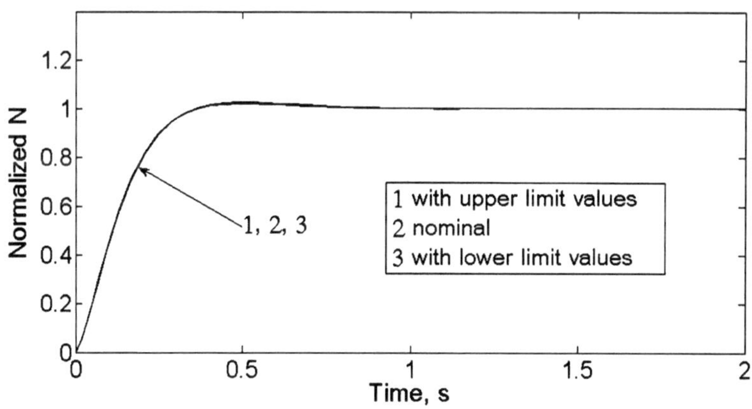

Fig. 3.15 Normalized speed as a function of time with disturbance in the presence of parametric uncertainties. Source: (Teh and Nassirharand 2010a) © Sage

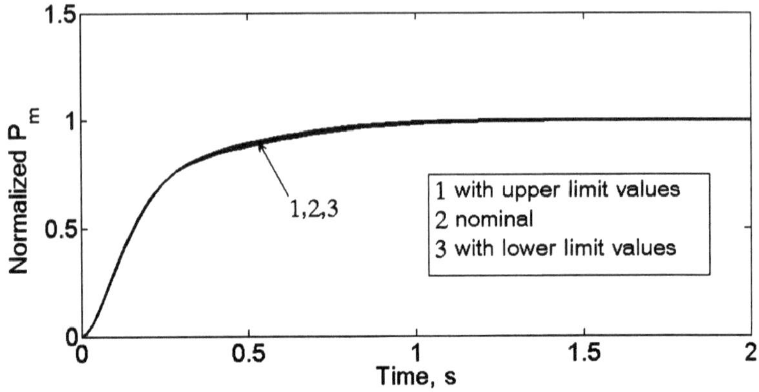

Fig. 3.16 Normalized manifold pressure as a function of time with disturbance in the presence of parametric uncertainties. Source: (Teh and Nassirharand 2010a) © Sage

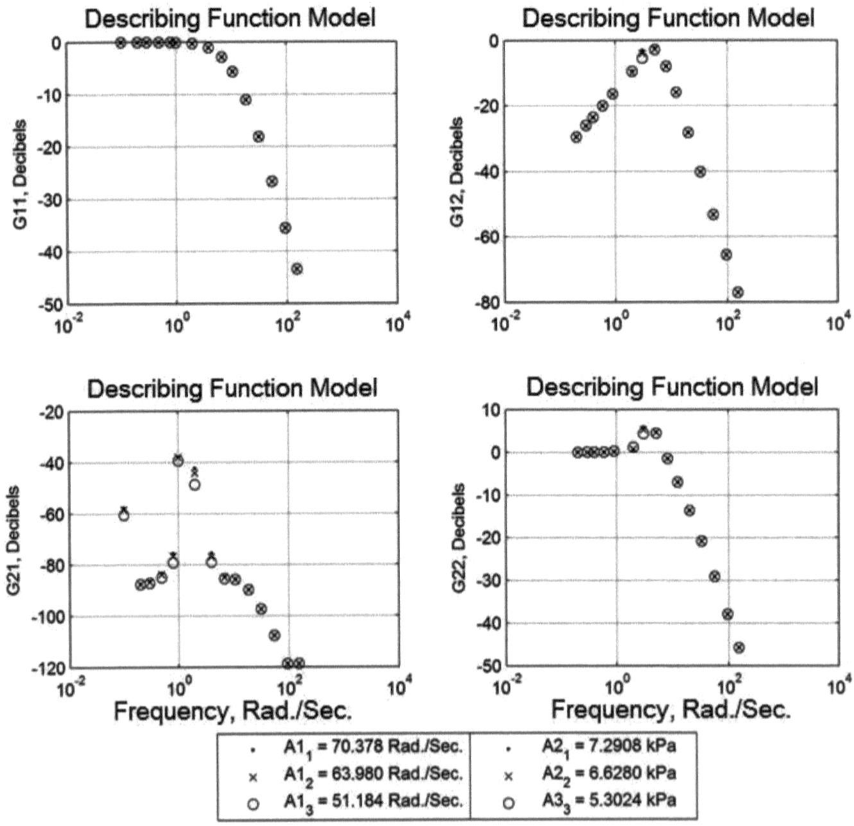

Fig. 3.17 Stability test – gain matrix of the closed-loop system (*A1* is speed and *A2* is manifold pressure). Source: (Teh and Nassirharand 2010a) © Sage

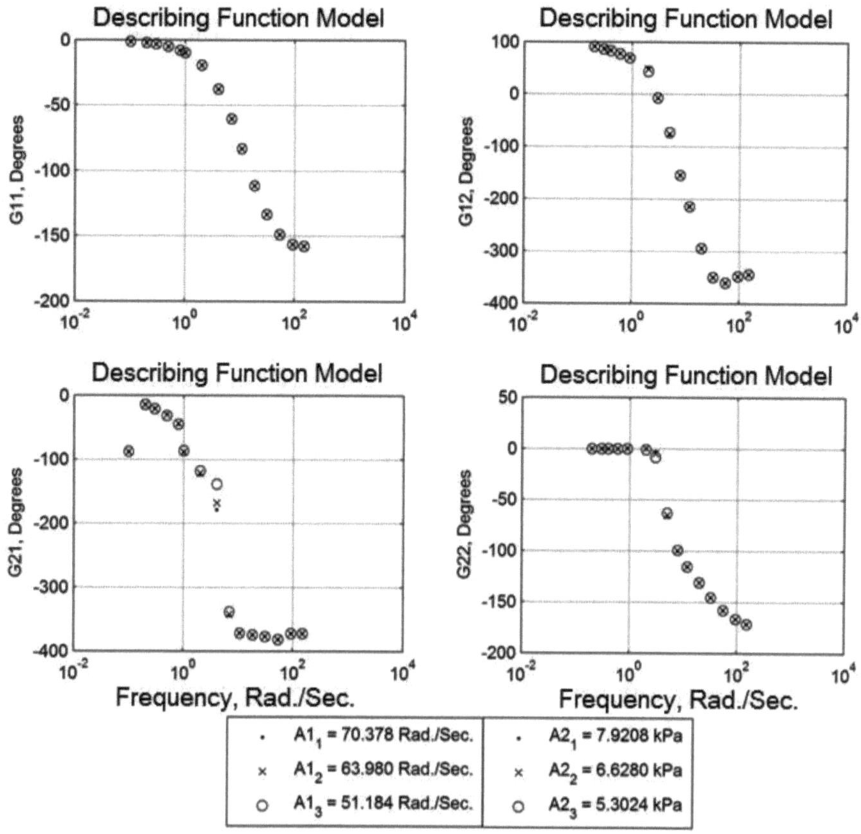

Fig. 3.18 Stability test – phase matrix of the closed-loop system (*A1* is speed and *A2* is manifold pressure). Source: (Teh and Nassirharand 2010a) © Sage

Stability of the final nonlinear closed-loop system is verified by successful generation of the corresponding SIDF models at different operating regimes. Stability and insensitivity to amplitude of excitation of the closed-loop system is examined by generating the closed-loop SIDF models at various command speeds and manifold pressures. These results are shown in Figs. 3.17 and 3.18 (Teh and Nassirharand 2010a). It is apparent that closed-loop system is stable and robust with respect to magnitude of the command signals.

Chapter 4
Dual-Range Controller Design

4.1 Background

In the early 1980s, the following question was considered. When is it really required to design a nonlinear controller for a highly nonlinear system? The need for a reply to this question received some attention (Taylor 1983). With the advancement of numerous robust control techniques for linear and nonlinear systems, the need to have a systematic procedure to answer the above question became even more evident. Taylor (1983) introduced the idea for design of dual-range linear controllers. In that work, he hypothesized that if one uses two quasi-linear models for controller design purposes, then controllers with a higher performance than those of their single-range counterparts may be designed. Consequently, it was suggested that if a dual-range linear controller did not provide adequate level of robustness, then a nonlinear controller design procedure should be considered. This hypothesis was verified for highly nonlinear single-variable systems (Nassirharand et al. 1988; Nassirharand 1991).

The original proposed approach for design of dual-range linear controllers for nonlinear systems called for application of the simultaneous stabilization theory of Vidyasagar and Viswanadham (1982). In that approach, one would parameterize the class of all stabilizing controllers in terms of a (matrix) transfer function parameter $R(s)$ followed by a search for the optimum controller. One iterative procedure for single-variable systems is proposed in Garvin and Mathew (1996). It has been discovered that searching for $R(s)$ results in high-order controllers and a direct search for the controller results in appreciably lower-order controllers (Nassirharand et al. 1988; Nassirharand 1991; Teh 2010). For example, when searching for $R(s)$ for a nonlinear plant with two integrators, a fourth-order controller is designed; a direct search results in a first-order simultaneously stabilizing controller (see Sect. 4.4). Furthermore, with this approach there is no guarantee that the controller that stabilizes two linear approximations of the nonlinear plant would also stabilize the actual nonlinear system. For this reason, in the new algorithm for design of dual-range controllers, there is a step that calls for tuning of the simultaneously

A. Nassirharand, *Computer-aided Nonlinear Control System Design: Using Describing Function Models*, DOI 10.1007/978-1-4471-2149-7_4,
© Springer-Verlag London Limited 2012

stabilizing controller. Examine Appendix D for a demonstration example problem that highlights the controller high-order problem as well as the stability problem mentioned above. It could be argued that appropriate coprime factorization could also result in competitive results; however, that would require further research which may not be justified under the conditions that a direct search for the simultaneously stabilizing controller is shown to be more straightforward.

It is also discovered that, when using the factorization theory, a high-order controller (i.e., of the order of 50) is required to stabilize only one of the plants given in the example problem of Sect. 4.5 (Teh 2010); furthermore, for more complex problems (such as quad-rotor helicopter problem (Al-Kindi and Nassirharand 2010)), MATLAB runs to numerical overflow problems. For this reason, it is recommended not to use the simultaneous stabilization theory of Vidyasagar and Viswanadham (1982) for applications to nonlinear systems. The preferred approach is to select a controller of a relatively low order (e.g., 0–4), with arbitrary coefficients that at least stabilizes one of the linear approximations, as a starting solution; then, optimization would be used to determine the coefficients of the controller. See Appendix D for justifications.

4.2 Controller Synthesis

The following algorithm is developed to obtain a desired dual-range linear controller (Teh and Nassirharand 2010b).

Steps 1–3 are similar to the single-range controller design discussed in the previous chapter. Step 4 is to select any controller that results in a stable closed-loop system for any one of the linear plants of step 3. For the purposes of this task, any linear design approach would suffice. In step 5, the *fminsearch* function of MATLAB® (which uses a Simplex (Reklaitis et al. 1983; Nelder and Mead 1965) search approach) may be utilized, and (for the cases that decoupling is not required) the following objective function may be minimized (Nassirharand et al. 1988; Patwardhan 1988; Nassirharand 1991; Teh 2010):

$$F = \alpha F_0 + (1-\alpha) F_1, \tag{4.1}$$

where

$$F_i = \beta\left[\left(y^i_{\max} - y^{d,i}_{\max}\right) + < y^i_{\min} >^2\right] + \sum_j \left(y^i_j - y^{d,i}_j\right)^2, \tag{4.2}$$

$$< y^i_{\min} >^2 = \begin{cases} \left(y^i_{\min}\right)^2 & \text{if } y^i_{\min} < 0 \\ 0 & \text{if } y^i_{\min} \geq 0 \end{cases}, \tag{4.3}$$

Algorithm 4.1

1. Derive the mathematical model of a reference linear model that the final closed-loop system is to mimic its dynamic and static behavior.
2. Obtain the describing function models of the nonlinear plant at various operating regimes (see Chap. 2).
3. Select two of the models of the previous step, and obtain linear fits.
4. Arbitrarily select a controller (of a low order, e.g., 0–4) with arbitrary coefficients that preferably stabilize at least one of the linear fits of the previous step.
5. Apply optimization, and optimize the coefficients of the dual-range linear controller for the two linear plants of step 3.
6. Apply optimization and tune the gains of the linear controller when placed to operate with the actual nonlinear plant.
7. Verify design of the final nonlinear closed-loop system. If satisfactory results are not obtained, proceed to design a nonlinear controller as outlined in Chap. 5; otherwise, stop.

d = a superscript indicating the desired value(s),
F = the objective function,
i = a superscript index (0 for nominal condition and 1 for the other),
j = a subscript index for unit step-response array,
max = a subscript indicating the maximum value,
min = a subscript indicating the minimum value,
y = the discretized closed-loop unit step response,
α = a weighting factor, $0 \leq \alpha \leq 1$,
β = a weighting factor, $\beta \geq 0$.

In Eq. 4.2, the terms in brackets avoid designs with excessive amounts of overshoot and undershoot.

In cases that decoupling is required, objective function (4.1) with the following alternative definitions is minimized (Nassirharand and Taylor 1990; Mousavi-Firdeh et al. 2006; Nassirharand and Karimi 2004b; Teh and Nassirharand 2010c; Teh 2010):

$$F_i = \left\{ \sum_{j'} \left(y^i_{j'} - y^{d,i}_{j'} \right)^2 + \sum_{k} \left(y^i_k \right)^2 \right\}, \tag{4.4}$$

where $j' = 1, 2, ..., l$, $k = 1, 2, ..., l$; $k \neq j'$, l = the number of outputs.

The rest of the parameters and variables are defined as before. It should be noted that one alternatively could modify Eq. 4.4 to penalize excessive amounts of overshoot and undershoot similar to Eq. 4.2.

In step 6, the controller gains are tuned by the same optimization routine as that used in the previous step. The experience of the authors in designing controllers for nonlinear plants based on a fitted transfer function (to the corresponding SIDF data) reveals that further tuning of the controller gains may be required before the controller could perform well when connected to the actual nonlinear plant. For this reason, the controller of the previous step is used as a starting solution, and a similar objective function to that presented by Eq. 4.1 is minimized in order to tune the controller parameters. In this case, y would be the normalized discretized closed-loop step response of the actual nonlinear plant when connected to the controller; also, index i would refer to the particular operating regime of interest.

Step 7 is to verify design. For stability, the following argument is used. If the describing function models of the final closed-loop system at various operating regimes are obtainable, then it may be concluded that Fourier integrals (see Eqs. 2.16 and 2.23) have converged; consequently, this suggests that the outputs must have been bounded. Hence BIBO (bounded-input bounded-output) stability may be demonstrated in this fashion. For verification of the performance of the closed-loop system, normalized step-response plots by exciting the plant with step inputs of various magnitudes may be obtained.

4.3 Software

A MATLAB® function is available that automates the design in step 5; the primary inputs to this function are the two linear plants identified in step 3 of Algorithm 4.1 as well as the reference linear model identified in step 1; the primary output is the dual-range controller. Another MATLAB® function is available that uses the Simulink model of the nonlinear plant and tunes the controller parameters obtained in step 5 (see Teh 2010 for details).

4.4 Example 1: A Single-Variable Servo System

Consider the SISO servo system that was also considered in Sect. 2.2.1.2. It is desired for the closed-loop system to mimic a second-order system with a damping ratio of 0.307 and a natural frequency of 43.22; similar performance measures were also achieved by Taylor and Strobel (1985) when designing a nonlinear PID controller. Algorithm 4.1 is executed as follows:

1. The reference linear model follows:

$$G_r(s) = \frac{\omega_n^2}{s^2 + 2\varsigma\omega_n s + \omega_n^2}, \tag{4.5}$$

where $\omega_n = 43.22\ rad/s$ and $\varsigma = 0.307$.

2. The describing function models are generated as outlined in Sect. 2.2.1. See Figs. 2.5, 2.6, 2.7 and 2.8 for pseudofrequency response plots.
3. The two linear models corresponding to small and large excitation amplitudes are selected. It is recommended to use the two linear models whose gain characteristics bound those of all the others in class; this corresponds to describing function models with excitation amplitudes of 0.25 V and 0.8 V. A linear fit is obtained, and the two linear models follow (Nassirharand et al. 1988; Nassirharand 1988):

$$G_0(s) = \frac{0.003989s + 1.231}{0.003761s^2 + 0.05475s}, \tag{4.6*}$$

$$G_1(s) = \frac{0.040314s + 1}{0.23641s^2 + 0.22521s}. \tag{4.7*}$$

4. The following simultaneously stabilizing (SS) controller is selected as the starting solution for the optimization:

$$C_{ss}^{start}(s) = \frac{1\,(s + 2)}{s + 4}. \tag{4.8}$$

5. The MATLAB® function DRLCD, which is outlined in Sect. 4.3, is utilized, and objective function (4.1) with Eq. 4.2 and setting $\alpha = 0.1$ and $\beta = 0$ is minimized; the following controller is obtained:

$$C_{ss}^{untuned}(s) = \frac{3.1671s + 1.1469}{0.0229s + 0.1089}. \tag{4.9}$$

6. The controller parameters of the previous step are tuned by optimization. The final synthesized controller is of the following form:

$$C_{ss}^{tuned}(s) = \frac{0.0301\,(1.1814s + 51.660)}{0.0017s + 0.1996}. \tag{4.10}$$

7. Design validation consists of generating the normalized closed-loop responses for different step magnitudes as well as generating the describing function models of the final nonlinear closed-loop system at various operating regimes. The latter is executed to demonstrate stability. The results are depicted in Figs. 4.1 and 4.2. The normalized step-response plots with a nonlinear PID controller are depicted in Fig. 5.6 (see Chap. 5 and Taylor and Strobel 1985). It is evident that the performance of the dual-range linear controller competes with a nonlinear PID controller.

---

*Source: (Nassirharand et al. 1988) © ASME

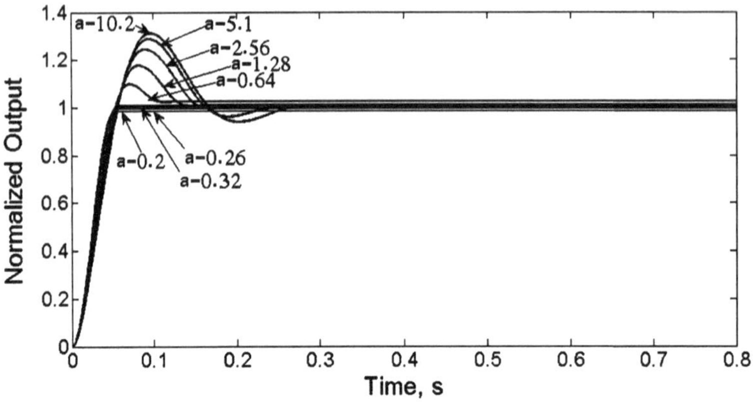

Fig. 4.1 Normalized step-response plots of the final closed-loop system with the dual-range linear controller

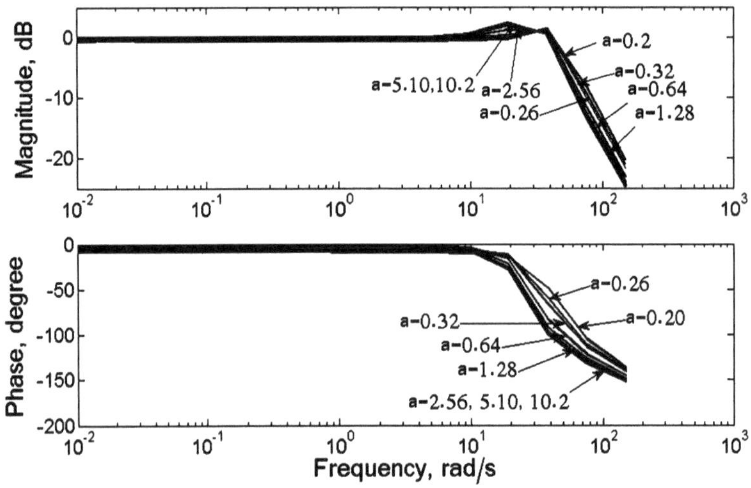

Fig. 4.2 Stability test – the generated describing function models of the final closed-loop system with the dual-range linear controller

4.5 Example 2: Multivariable Bank Angle Control Problem

Consider the unmanned aerial vehicle bank angle control problem that is studied in Colgren 2004; the corresponding schematic block diagram is depicted in Fig. 4.3. The problem statement is to design a dual-range linear controller for this nonlinear system that would force the bank angle response of the corresponding nonlinear feedback system to mimic a second-order system with a damping ratio of 0.9 and a natural frequency of 1.11 rad/s; these are approximately the same performance

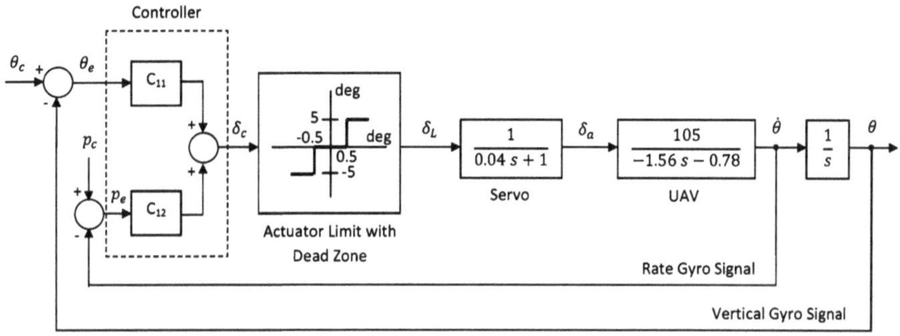

Fig. 4.3 The schematic diagram of the UAV and the structure of the closed-loop system. Source: (Teh 2010) © S.H. Teh

measures that are achieved when designing a multivariable nonlinear controller as outlined in Chap. 5 (Nassirharand 2010a). The details of the application of the design procedure given in the previous section follow (Teh 2010):

1. With reference to Fig. 4.3 and the problem statement, the following reference linear model is considered:

$$G_r(s) = \begin{bmatrix} \dfrac{\omega_n^2 s}{s^2 + 2\varsigma\omega_n s + \omega_n^2} & 0 \\ 0 & \dfrac{\omega_n^2}{s^2 + 2\varsigma\omega_n s + \omega_n^2} \end{bmatrix}, \tag{4.11*}$$

   where $\omega_n = 1.11\text{rad/s}$ and $\varsigma = 0.9$.
2. In this step, the input/output behavior of the nonlinear plant at various operating regimes is to be characterized. The sinusoidal-input describing function models of the nonlinear plant are obtained for eight input amplitude values in the frequency range of 0.1–1,000 rad/s. The input amplitude values considered were 0.75°, 1°, 2°, 4°, 8°, 16°, 32°, 64°, and 128°. The resulting frequency response plots are depicted in Fig. 4.4.
3. The two quasi-linear models corresponding to input amplitude levels of 1° and 8° are arbitrarily selected for system identification purpose. The MATLAB® function *invfreqs* is utilized for this purpose. The identified models are given by the following equations; the quality of fit may be examined by considering Figs. 4.5 and 4.6.

$$G_0(s) = \begin{bmatrix} \dfrac{-70200}{s^2 + 23.08s + 96.81} \\ \dfrac{-70200}{s\,(s^2 + 23.08s + 96.81)} \end{bmatrix}, \tag{4.12*}$$

*Source: (Teh 2010) © S.H. Teh

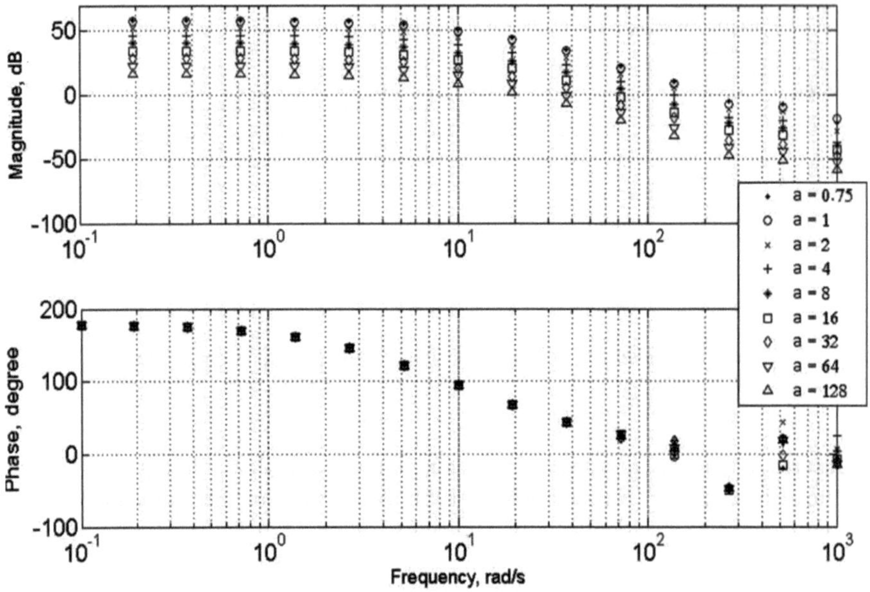

Fig. 4.4 UAV open-loop pseudofrequency response plots. Source: (Teh 2010) © S.H. Teh

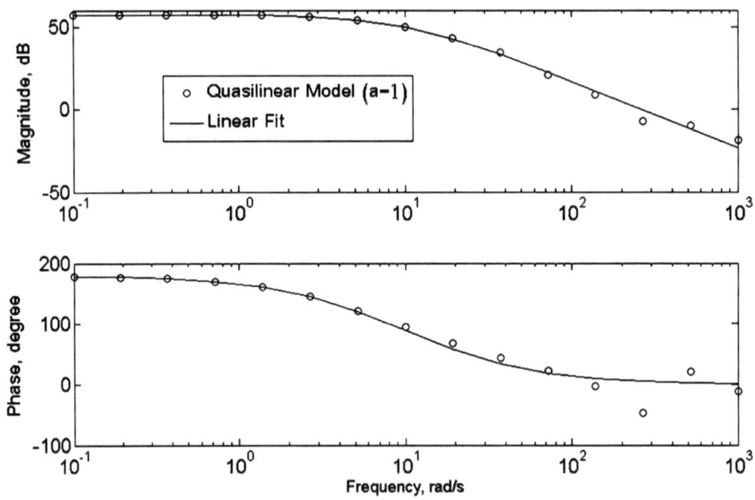

Fig. 4.5 Linear system identification at a low excitation amplitude. Source: (Teh 2010) © S.H. Teh

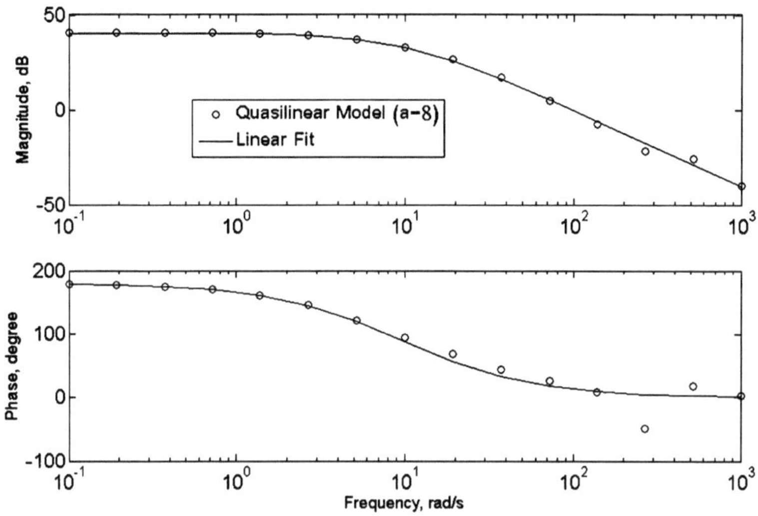

Fig. 4.6 Linear system identification at a high excitation amplitude. Source: (Teh 2010) © S.H. Teh

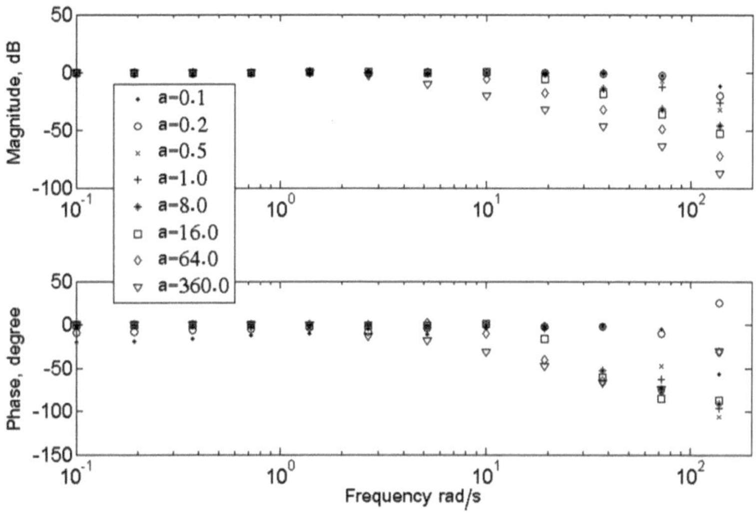

Fig. 4.7 Stability test. Source: (Teh 2010) © S.H. Teh

$$G_1(s) = \begin{bmatrix} \dfrac{-9606}{s^2 + 22.15s + 92.54} \\ \dfrac{-9606}{s\,(s^2 + 22.15s + 92.54)} \end{bmatrix}. \tag{4.13*}$$

*Source: (Teh 2010) © S.H. Teh

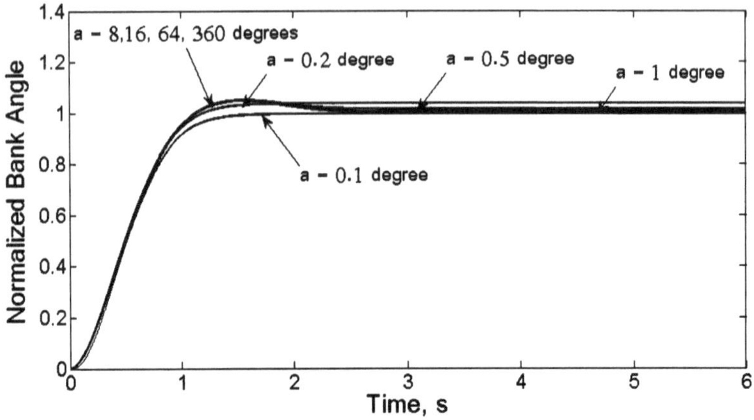

Fig. 4.8 Normalized step-response plots of the synthesized dual-range linear controller and the nonlinear UAV. Source: (Teh 2010) © S.H. Teh

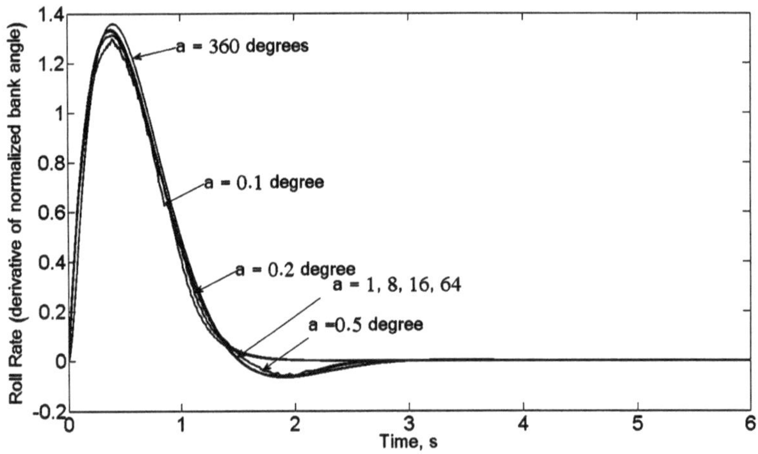

Fig. 4.9 Normalized roll-rate plots of the synthesized dual-range linear controller and the nonlinear UAV. Source: (Teh 2010) © S.H. Teh

4. Arbitrarily, select the parameters of the starting solution of the controller. In order to appreciate the developments in this chapter, a starting solution using the application of simultaneous stabilization theory is obtained as detailed in Teh 2010. It is not necessarily recommended to use this approach to obtain a starting solution, and it is given here for illustration purposes only. The following starting solution for the controller is selected:

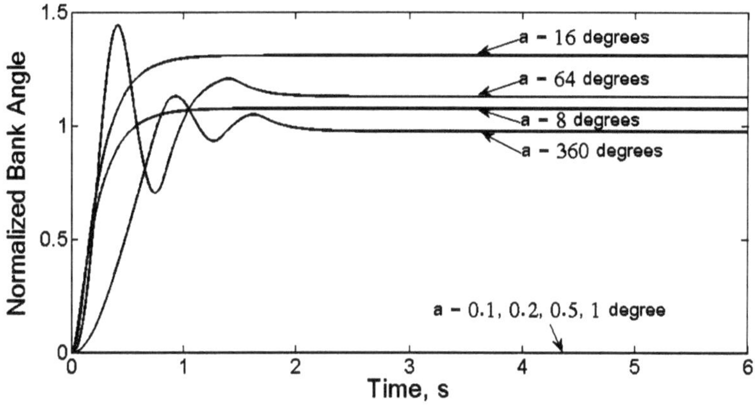

Fig. 4.10 The normalized step-response plots of the closed-loop system with a single-range linear controller and the nonlinear UAV. Source: (Teh 2010) © S.H. Teh

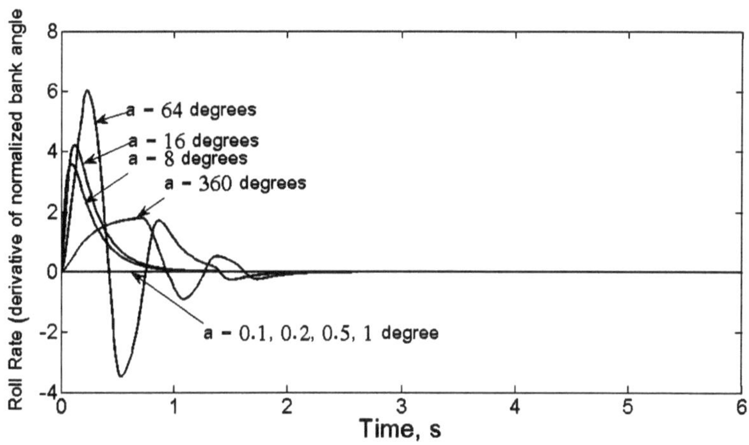

Fig. 4.11 The normalized roll-rate plots of the closed-loop system with a single-range linear controller and the nonlinear UAV. Source: (Teh 2010) © S.H. Teh

$$C^{\text{start}}(s) = \left[\frac{995.4\left(s^2 + 5.8s + 147.3\right)}{s^2 + 43751.9s + 8.8E6} \;\; \frac{4.6\left(s^2 + 16270.9s - 474819\right)}{s^2 + 43751.9s + 8.8E6} \right]. \tag{4.14*}$$

5. The MATLAB® function described in Section 4.3 is used, and objective function (4.1) with Eq. 4.2 is minimized to obtain the following simultaneously stabilizing controller:

*Source: (Teh 2010) © S.H. Teh

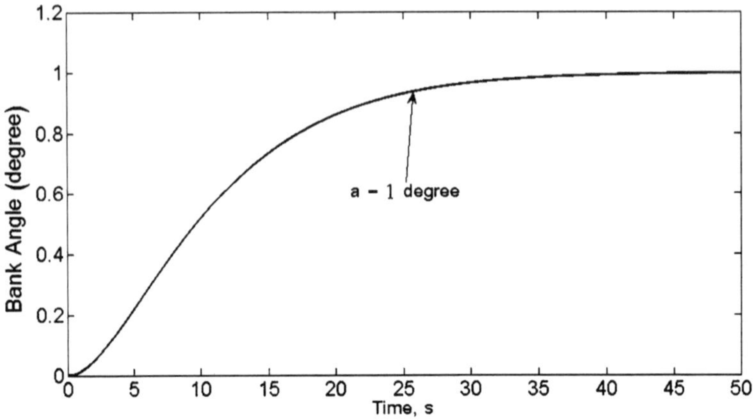

Fig. 4.12 The step-response plot of a nonlinear H_∞ controller and the nonlinear UAV. Source: (Colgren 2004) © R. Colgren

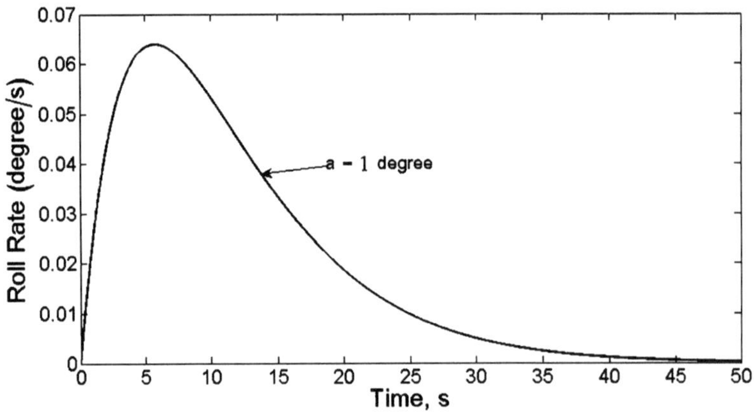

Fig. 4.13 The roll-rate plot of a nonlinear H_∞ controller and the nonlinear UAV. Source: (Teh 2010) © S.H. Teh

$$C_{ss}^{\text{untuned}}(s) = \begin{bmatrix} \dfrac{-0.48s^2 - 1.94s + 2.45E - 4}{0.67s^2 + 7.20E - 4s} \\ \dfrac{-2.86E - 4s^2 - 0.40s + 4.36E - 5}{0.11s^2 + 1.24E - 4s} \end{bmatrix}^T. \quad (4.15^*)$$

*Source: (Teh 2010) © S.H. Teh

6. The parameters of the controller in Eq. 4.15 are tuned as outlined in Sect. 4.2.

$$C_{ss}^{\text{tuned}}(s) = \begin{bmatrix} \dfrac{-1.50\left(0.72s^2 + 3.0s + 6.61E-4\right)}{0.26s^2 + 5.5E-4s} \\ \dfrac{-15.53\left(2.62E-4s^2 + 0.30s + 5.1E-5\right)}{0.21s^2 + 1.01E-4s} \end{bmatrix}^T. \qquad (4.16^*)$$

7. The describing function models of the closed-loop system at various operating regimes are depicted in Fig. 4.7. It is apparent that the system is stable. For verification of performance measures, the normalized step-response plots of the bank angle and roll-rate are given in Figs. 4.8 and 4.9, respectively. The normalized step-response plots of the bank angle and roll-rate of a single-range linear controller (Nassirharand and Mousavi-Firdeh 2009b), the normalized step-response plots of the bank angle and roll-rate of a multi-range nonlinear controller (Nassirharand 2010a), and one step-response plot of the bank angle and roll-rate of a nonlinear H_∞ controller (Colgren 2004) are shown in Figs. 4.10, 4.11, 4.12, 4.13, 5.20 and 5.21, respectively. It is apparent that the performance of the dual-range linear controller is superior to that of the single-range linear controller, and it competes with the two other nonlinear controllers.

*Source: (Teh 2010) © S.H. Teh

Chapter 5
Multirange Nonlinear Controller Design

5.1 Background

The first attempt at design of nonlinear PID controllers based on sinusoidal-input describing function models and nonlinearity inversion using inverse describing function technique is reported in Taylor and Strobel (1985) and later in Nassirharand and Karimi (2006). Competitive work (using random-input describing functions) is reported in Suzuki and Hedrick (1985) as well as Nanka-Bruce and Atherton (1990). A series of MATLAB® functions is also developed in Nassirharand and Karimi (2006) that automates the design procedure of Taylor and Strobel (1985). Following this, design of nonlinear lead-lag and nonlinear H_∞ controllers is developed in Nassirharand and Mousavi Firdeh (2008) and in Nassirharand and Mousavi Firdeh (2009c), respectively. See Nanka-Bruce and Atherton (1990) for competitive nonlinear lead-lag controller design, and see Colgren (2004) for an alternate nonlinear H_∞ controller design.

5.2 Controller Synthesis

The following algorithm may be utilized to obtain a desired nonlinear controller for a highly nonlinear plant characterized by Eqs. 1.1 and 1.2 (Taylor 1983; Nassirharand and Karimi 2006; Nassirharand and Mousavi Firdeh 2008, 2009c).

Steps 1 and 2 are similar to the single-range controller design discussed in Chap. 1. In step 3, any linear controller design technique may be used to design a linear controller. For example, one could fit a linear model to the nominal pseudofrequency response data followed by application of the *cad_controller* software (Nassirharand 2009b) to design a PID (Nassirharand et al. 2003.), lead-lag (Nassirharand and Karimi 2004c), or a *n*th-order transfer function (Nassirharand 2008a). Also, MATLAB® may be used and a linear H_∞ controller may be designed. Furthermore, it is recommended to check the step response of the designed linear controller

A. Nassirharand, *Computer-aided Nonlinear Control System Design: Using Describing Function Models*, DOI 10.1007/978-1-4471-2149-7_5,
© Springer-Verlag London Limited 2012

and the actual nonlinear plant at nominal conditions before proceeding to the next step. The parameters of the linear controller may have to be tuned similar to that described in step 6 of Algorithm 4.1. Step 4 is self-explanatory. In step 5, in order to compute Eq. 5.1, one must place the describing function generator software inside the optimization loop. This will cause the execution time of the simulation code to unreasonably increase, and design for industrial systems with the usual accessible computers becomes almost impossible. In order to lift this restriction, to compute CG_i, one could compute $C_i\,(j\omega)\,.G\,(j\omega, e_i\,|C_i\,(j\omega)|)$, where

$C_i =$ the controller at operating regime i
$e_i =$ amplitude of the sinusoid at the input to the controller
$G =$ the pseudofrequency domain model of nonlinear plant at a specific operating regime
$j = \sqrt{-1}$
$\omega =$ frequency, rad/s

To accomplish this, the $G\,(j\omega, a_i)$ data obtained in step 2 coupled with interpolation is utilized. This computation method considerably increases the execution speed of step 5, and minimizing $E\,(j\omega)$ via the *fminsearch* function of MATLAB® for complicated industrial systems via standard computers may easily be accomplished. Step 6 is described in the next subsection, and in step 7, normalized step-response plots with various step magnitudes of interest are obtained. For demonstration of stability, the describing function models of the final closed-loop system at various operating regimes may be generated.

Algorithm 5.1

1. Derive the mathematical model of a reference linear model that the final closed-loop system is to mimic its dynamic and static behavior.
2. Obtain the describing function models of the nonlinear plant at various operating regimes (see Chap. 2).
3. Select one of the input-output models of the previous step as the nominal model, and design a linear controller for this nominal model. The structure of the linear controller could be a PID, a lead-lag, a H_∞ controller, or simply a nth-order transfer function.
4. Place the model of the designed controller of the previous step in series with the nonlinear plant. Using the technique outlined in Chap. 2, generate the corresponding open-loop describing function model. This generated open-loop model is the desired model that the open-loop system is to mimic at all operating regimes. It should be noted that in this step, the excitation signal is obtained by dividing the amplitude which corresponds to the selected nominal model used in step 3 by the gain of the linear

controller at a specific frequency. For example, the controller gain at the phase crossover frequency may be used (Taylor and Strobel 1985); for a type zero controller like a lead-lag controller or a controller that does not have a crossover frequency, the low-frequency gain may be used (Nassirharand and Mousavi Firdeh 2008, 2009c).

5. In this step, a set of linear controllers are designed such that the following objective function over the frequency set of interest is a minimum.

$$E\,(j\omega) = |C^*G^* - C\,G_i| \tag{5.1}$$

where

$C\,G_i$ = the pseudofrequency response model of the controller and the nonlinear plant at operating regime i

C^*G^* = the desired pseudofrequency response model obtained in step 4

E = a measure of the error between the desired open-loop pseudofrequency response model obtained in step 4 and the open-loop pseudofrequency response model comprised of the controller and the nonlinear plant at operating condition i

It should be noted that $E\,(j\omega)$ measures the sensitivity (or robustness) of the open-loop system, and minimizing $E\,(j\omega)$ at various operating regimes of interest results in the required amplitude-dependent behavior of a minimum sensitivity controller. The output of this step is a set of controller parameters for each e_i.

6. Interpret the table of values for each controller parameter as a function of e_i (which was obtained in the previous step) as the describing function model for that controller parameter; by using the SIDF model inversion technique described below (Nassirharand 2009a), invert the describing function model for each controller parameter to identify the desired nonlinear function for that controller parameter.
7. In this step, a model of the nonlinear controller which incorporates the synthesized nonlinear functions of the previous step is constructed, and the design is verified. For verification, the describing function models of the synthesized nonlinear closed-loop system may be generated to demonstrate stability; in other words, if the Fourier integrals (see Eqs. 2.16 and 2.23) converge at all operating regimes, then stability may be concluded. Furthermore, the normalized step-response plots with various magnitudes of command signal may also be generated to verify time-domain behavior. Additionally, in order to quantify the amount of achieved reduction in sensitivity, the followings are evaluated numerically:

$$\sigma_0 = \sum_i \left[\int_{\omega_1}^{\omega_2} \left| 1 - \frac{G_i\,(j\omega, a_i)}{G^*\,(j\omega, a^*)} \right| \mathrm{d}\omega \right] \tag{5.2}$$

$$\sigma_1 = \sum_i \left[\int_{\omega_1}^{\omega_2} \left| 1 - \frac{\Gamma_i}{\Gamma^*} \right| \mathrm{d}\omega \right] \tag{5.3}$$

where

a_i = amplitude of the sinusoid at the input to the nonlinear plant at operating regime i
a^* = amplitude of the sinusoid at the input to the nonlinear plant at nominal conditions
Γ_i = the pseudofrequency domain model of the open-loop system comprised of the designed nonlinear controller and the actual nonlinear plant at operating regime i
Γ^* = the pseudofrequency domain model of the open-loop system comprised of the designed linear controller and the actual nonlinear plant at nominal conditions (see step 3)
σ_0 = measure of sensitivity of the nonlinear plant
σ_1 = measure of sensitivity of the designed open-loop system

The closer σ_k $(k = 0, 1)$ is to zero, the smaller is the level of sensitivity. The percent amount of reduction in sensitivity would be given by the followings relation:

$$R = \frac{\sigma_0 - \sigma_1}{\sigma_0} \times 100 \tag{5.4}$$

5.2.1 *Describing Function Inversion (Nassirharand 2009a)**

The inversion procedure starts by guessing the initial values for the independent variables of an assumed nonlinear function. Then, the describing function model of the nonlinearity is numerically generated. This is followed by minimizing an objective function that yields the desired parameters of the nonlinear function.

Identification of linear systems from frequency response data is not a new subject, and there is considerable literature that covers this subject (e.g., see Nassirharand 1988 and references therein). Unlike nonlinear systems, this problem

*Source: (Nassirharand 2009a) © Elsevier

for linear systems is independent of amplitude. One application of identification of systems from frequency response data is in the area of controller design.

The focus of this section is on identification of nonlinear models or nonlinear functions from their corresponding describing function models. What is common in most identification works is that a model for the function/system has to be assumed, and then the parameters of that model are identified. This development is of no difference, and a model for the function has to be assumed whose parameters may be identified using a newly developed algorithm. For identification of nonlinear system structure, see Petrovic et al. (2000), Haber and Keviczky (1999), and Bongard and Lipson (2005).

Lyashevskiy and Chen (1997) and Besancon-Voda and Blaha (2002) used describing functions to identify the parameters of nonlinear functions of a nonlinear system. Those works are based on imposing harmonic balance conditions, which lead to a set of nonlinear algebraic equations that must be solved numerically. One problem associated with that approach is that input to each nonlinear term is assumed to be sinusoidal; such assumptions limit the application of the technique. In this work, simulation and evaluation of Fourier integrals are utilized, and there are no assumptions that input to each nonlinearity is sinusoidal. A method is also developed that uses describing function approach for expressing the nonlinearity in the structure (Ozer et al. 2009); then, using measured responses, one can calculate the describing function model and perform identification. Unlike this work, this requires one to isolate the nonlinear term within the system. A method for identification of friction using describing functions has also been developed by Kim and Chung (2006); unlike this work, the method of Kim and Chung (2006) is limited to identification of friction elements within servo systems. Prazenica et al. used Volterra theory to identify nonlinear system dynamics from measured input-output data. This approach is useful if the system includes polynomial-type nonlinear terms, and (unlike this work) it may not be applied to systems with discontinuous nonlinear terms. An algorithm for identification of a system divided into cascaded blocks of dynamic linear, static nonlinear, and dynamic linear subsystems has also been developed (Shi and Sun 1990). A Kernel-based version of recursive least-squares algorithm for nonlinear system identification purposes was developed by Vaerenbergh et al. (2007). A nonlinear receptance coupling approach (NRCA) was used to identify a nonlinear force function of a device for evaluation of machine dynamic characteristics (Dhupia et al. 2008). Multiple integration method was used to identify nonlinear single-variable systems from measured data (John 2001); note that in principle the developed algorithm is applicable to identification of multivariable systems as well. However, its application is left for future work. There are other nonlinear system identification techniques based on neural networks (Ge et al. 2008) and fuzzy theory (Li et al. 2010). See Kerschen et al. (2006) for a literature survey on the subject. The primary differences between this work and other noted literature are: (1) the procedure is computer-aided, (2) the procedure is not limited to inversion of memoryless nonlinear terms, and (3) the corresponding MATLAB® software is presented.

For static nonlinear terms the describing function model is simply the ratio of the fundamental output (obtainable by a Fourier analysis described in Chap. 2) to the amplitude of the applied sinusoidal input. The inverse problem considered here is simply one of being given the describing function and finding the corresponding nonlinearity. The analytical solution to this problem is given by Gelb and Vander Velde (1968) and Colgren (2004), and the interested reader is referred to that literature; the analytical solution will not be repeated here. The analytical solutions are good only for the class of considered nonlinear functions, and for a complete analytical solution, one should theoretically develop a catalogue of infinite analytical solutions. As an alternative approach, a numerical technique may be considered. In this approach, the user assumes the structure of the nonlinear function, and the model parameters are identified. Therefore, the form of the assumed nonlinear function may also have memory, and the fact that nonlinear function may have memory does not limit the application of the procedure.

The inversion process in step 6 of Algorithm 5.1, has received limited attention (Atherton 1970). At the first sight, it might appear that if a static output versus input graph consisting of piecewise linear segments is to be inverted, it can be accomplished readily by reordering the axis without the algorithm developed in this section. To clarify this matter, assume that in step 6 of Algorithm 5.1, the nonlinear function describing one of the controller gains as a function of the excitation amplitude is determined to be given by Fig. 2.2. As was mentioned, the nonlinear function in Fig. 2.2 is interpreted as the describing function model of an as yet unknown nonlinear function; this unknown function describes the controller gain as a function of the error or the excitation signal. This nonlinear function is given by Fig. 2.1. In other words, Fig. 2.1 is the inversion of describing function model given by Fig. 2.2. In step 6 of Algorithm 5.1, Fig. 2.1 must be obtained by inversion of the describing function model given in Fig. 2.2. It is clear that reordering the axes, without the algorithms developed in this section, would not result in the desired nonlinear function. Furthermore, unlike inversion of Fig. 2.2, it is generally not easy to obtain the inverted format as noted in Atherton (1970) and research in Nassirharand and Mousavi Firdeh (2008, 2009c), Taylor and Strobel (1985), and Nassirharand and Karimi (2006). In Gelb and Vander Velde (1968), early need for inversion of describing function models was noted. Then Atherton (1970) and Colgren (2004) gave directions and procedures for inversion of memoryless describing function models.

The procedure for inversion of describing function models is computer-aided. Similar to other available inversion methods, the user must select a model for the nonlinear function. Previous works demonstrate that a general piecewise linear function appears to be a good starting point (Taylor 1983; Taylor and Strobel 1985; Nassirharand and Karimi 2006; Nassirharand and Mousavi Firdeh 2008, 2009c). This choice has the following features:

1. Allows linear gain at small amplitudes where the plants usually behaves as a linear system
2. Allows sudden jumps or reduction in controller gain at intermediate operating regimes
3. Allows gradual increase or reduction of gains at large amplitudes

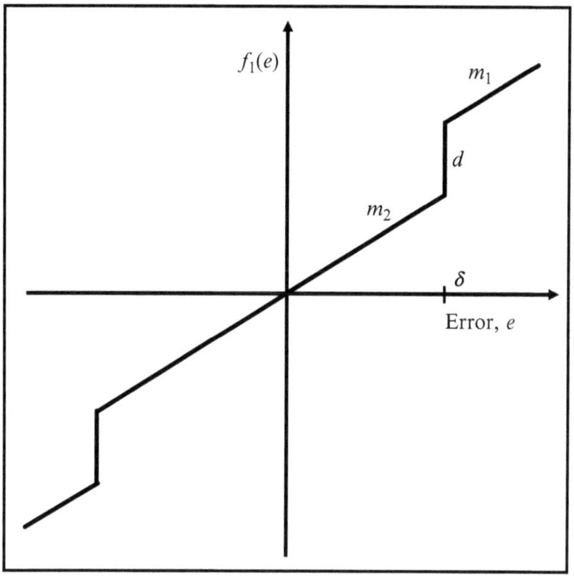

Fig. 5.1 A simple candidate nonlinear function for inversion (Taylor 1983)

The designer may also choose to examine the catalogue of nonlinear functions given in the "Appendix" of Gelb and Vander Velde (1968), and select the model that seems to suite his problem. The approach is independent of the selected model. Once a model is selected, it must be described in terms of its parameters as a MATLAB® function. Then, a set of initial values for parameters of this model are selected, and the describing function model of the assumed nonlinear function is digitally obtained (Nassirharand and Karimi 2002). Finally, an objective function measuring the error between the desired describing function model and that of the actual nonlinear function is minimized. This minimization is done by using *fminsearch* function of the MATLAB® software. The developed algorithm for describing function inversion is given below.

In step 1, the user must define a model for the nonlinearity. To accomplish this task, the user could plot the given describing function model. Then, he could compare this plot with the catalogue of describing function models given in "Appendix" of Gelb and Vander Velde (1968). Finally, he would select the model that may exhibit the desired gain/phase-amplitude behavior. The state of art for controller synthesis applications suggests a nonlinear function similar to those in Figs. 5.1, 5.2, and 5.3.

Maybe in future, nonlinear gain functions with memory will be needed. In step 2, only a guess is required. Again for controller synthesis applications, by looking at the desired gain-amplitude plot, one could easily determine an initial guess for the parameters. In step 3, the MATLAB function that is given in Chap. 2 is modified to fit SISO models, and this function is used to obtain the describing function model of the assumed nonlinear function. In step 4, the objective function is evaluated, and

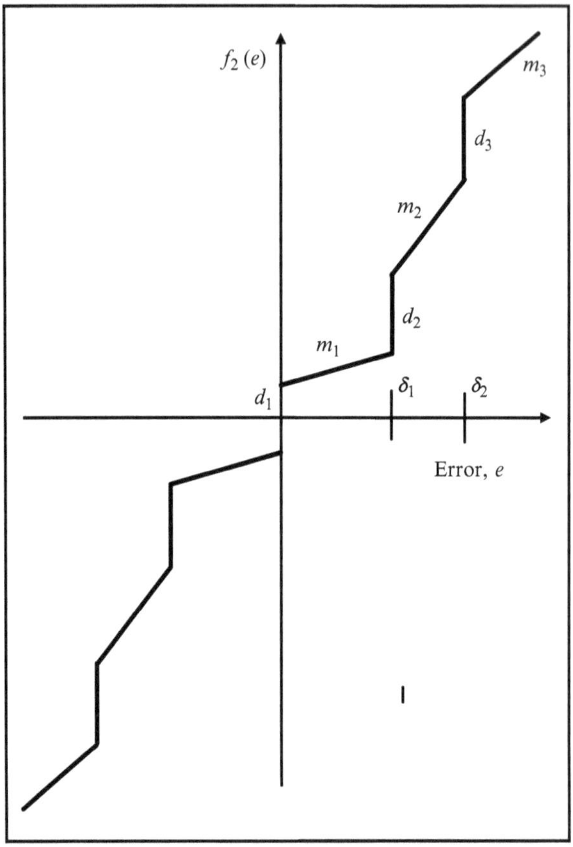

Fig. 5.2 A complicated nonlinear function with an abrupt change at origin for inversion (Nassirharand and Mousavi Firdeh 2008)

then minimized using an optimization routine. Specifically, the *fminsearch* function of the MATLAB® that uses a Simplex (Reklaitis et al. 1983; Nelder and Mead 1965) search technique is used here. In step 5, the inversion results are verified.

Algorithm 5.2

1. Assume a model for the desired actual nonlinear function.
2. Determine a set of initial values for the parameters of the model in step 1.
3. Use the *getsidfinv2* function developed here, and obtain the describing function (DF) model of the assumed nonlinear function.
4. Minimize the following objective function:

$$E = \sum_i |G_{i,D}(j\omega) - G_{i,A}(j\omega)|^2 \tag{5.5}$$

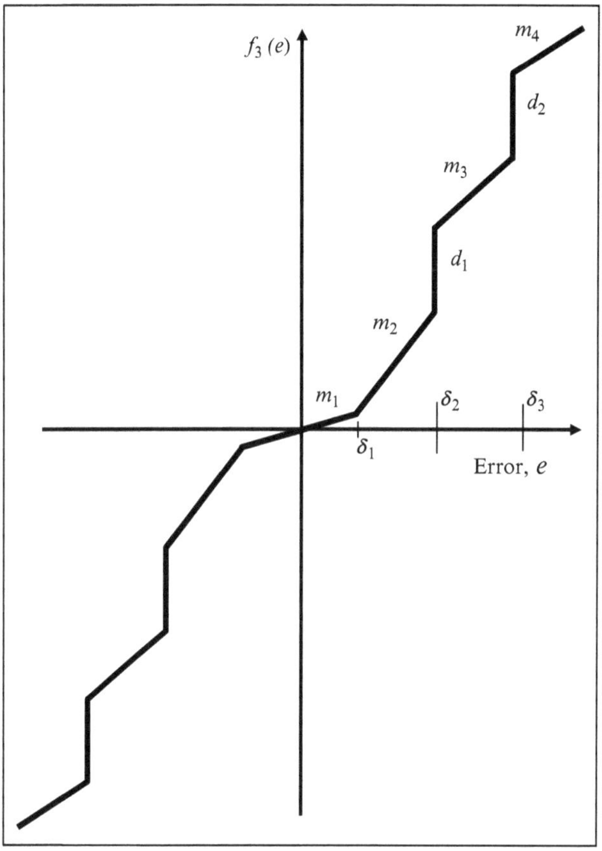

Fig. 5.3 A complicated nonlinear function with linear gain at origin for inversion (Nassirharand and Mousavi Firdeh 2009c)

where

$G_{i,D}(j\omega)$ is the given describing function model at operating regime i, $G_{i,A}(j\omega)$ is the actual describing function model, $\omega$is the frequency, and $j = \sqrt{-1}$.

If the assumed nonlinear function is memoryless, then Ewould be independent of frequency.

5. Verify the inversion results. If satisfactory results are obtained, then stop; otherwise, change the initial guess and/or the assumed model of step 1 and go to step 3.

5.3 Software

5.3.1 Software for Step 5

An example software for optimization of the gains of the controller (i.e., a PID controller) is presented in Fig. 5.4. The main program reads in the values of the pseudofrequency responses of the plant and the desired open-loop model; initial guesses for the controller gains are set, and finally the gains are optimized. Refer to the comments inside the required functions for details.

```
%
% This is the main program for optimizing the gains of the
% controller
% Data file cgstart includes the results of describing function
% generation of the plant (i.e., g(jw)) and the desired open-loop
% model (C*G*) from previous steps
load cgstart
global gjw amplg wg e cgstarg ejomega aux
% grt and git are the real and imaginary part of the desired open
% loop frequency domain model C*G*
%
cgstarg=complex(grt,git);
%ampl and w are the amplitude and frequency set used to generate the
% plant pseudo frequency domain model
amplg=ampl;wg=w;
sw=size(w);nw=sw(2);
sa=size(ampl);na=sa(2);
for ia=1:na
   for iw=1:nw
%
% gr and gi are the real and imaginary part of the frequency domain
% model of the nonlinear plant that are loaded in via the load
% function at the beginning of the program
      gjw(ia,iw)=complex(gr(ia,iw),gi(ia,iw));
   end
end
% x0 is the initial guess for Kp, Ki, and Kd
x0=[4 40 0.08];
%
%start optimizing the gains for each amplitude; there are eight
% amplitudes
for i=1:8
% obtain the amplitude of the error signal by dividing the amplitude
% at the input to the nonlinear plant by the gain at the cross-over
%
    e=amplg(i)/3.9811;
% use fminsearch and optimize the gains
    x=fminsearch('getejomega',x0);
% store the optimized values; ejomega is the value of the Objective
% Function (5.1)
    kp(i)=x(1);ki(i)=x(2);kd(i)=x(3);el(i)=e;error(i)=ejomega;
end
```

Fig. 5.4 Listing of the software for optimization of the gains

```
%
% this function calculates the value of Objective Function
% (5.1)
%
function ejomega=getejomega(x)
%x is the vector of independent variables or the gains Kp, Ki,
% and Kd
global wg e cgstarg ejomega
sw=size(wg);nw=sw(2);
% calculate the frequency response values of the controller
% with the present values of the gains and the assumed values
% of the frequency (wg)
cjw=getcjw(x,wg);
%
% calculate the pseudo frequency response values of the plant
% by interpolation; e is the amplitude of the input signal to
% the controller block and nw is the number of considered
% frequencies
gjwca=getgjwca(wg,e,cjw,nw,x);
% calculate the objective function
%
cejomega=ones(size(cjw))-(cjw.*gjwca)./(cgstarg);
ejomega=0;
for i=1:nw
   ejomega=ejomega+(abs(cejomega(i))^2);
end
```

```
% this function calculates the frequency response of the
% controller and it is self explanatory
function cjw=getcjw(x,w)
kp=x(1);ki=x(2);kd=x(3);
num=[kd kp ki];den=[1 0];
s=sqrt(-1)*w;c1=polyval(num,s);c2=polyval(den,s);
cjw=c1./c2;
%
%this function calculates the pseudo frequency response of the
% nonlinear plant by interpolation
function gjwca=getgjwca(w,e,cjw,nw,x)
global amplg
for iw=1:nw
% calculate the amplitude of the plant input signal
   a=e*abs(cjw(iw));
%calculate the two amplitudes, real-parts, and imaginary parts
% to be used for interpolation/extrapolation
   [a1,a2,rl1,rl2,im1,im2]=getintp(amplg,a,iw);
   rl=intp(a,a1,a2,rl1,rl2);im=intp(a,a1,a2,im1,im2);
   gjwca(iw)=complex(rl,im);
end
```

Fig. 5.4 (continued)

```
% this function calculates the data points to be used for
% interpolation/extrapolation
function [a1,a2,rl1,rl2,im1,im2]=getintp(ampl,a,iw);
% iw is the index of the frequency of interest
% gjw is the pseudo frequency response data of the
% nonlin  ear plant
% obtained in the previous Step
global gjw
sampl=size(ampl);iampl=sampl(2);
for i = 1:iampl
   if(a <= ampl(1))
% the amplitude of interest is below the lowest value available
       a1=ampl(1);a2=ampl(2);
       rl1=real(gjw(1,iw));rl2=real(gjw(2,iw));
       im1=imag(gjw(1,iw));im2=imag(gjw(2,iw));
       return
   elseif ( a >= ampl(iampl))
% the amplitude of interest is above the highest value available
       a1=ampl(iampl-1);a2=ampl(iampl);
       rl1=real(gjw((iampl-1),iw));rl2=real(gjw((iampl),iw));
       im1=imag(gjw((iampl-1),iw));im2=imag(gjw((iampl),iw));
       return
   end
end
for i=1:iampl
   if (ampl(i) >= a)
% store the data for interpolation
       a1=ampl(i-1);a2=ampl(i);
       rl1=real(gjw((i-1),iw));rl2=real(gjw((i),iw));
       im1=imag(gjw((i-1),iw));im2=imag(gjw((i),iw));
       return
   end
end
```

```
%
% this function performs linear
% interpolation/extrapolation and it is
% self-explanatory
function y=intp(x,x1,x2,y1,y2);
slope=(y2-y1)/(x2-x1);
y=slope*(x-x1)+y1;
%end
```

Fig. 5.4 (continued)

5.3.2 *Software for Describing Function Inversion (Nassirharand 2009a)*

An example software to achieve the inversion task is presented in Fig. 5.5. In this figure, a main program, the function that calculates the Objective Function (5.5) (i.e., *errorinv*), the function that generates the describing function model of the

```
%
% THIS IS THE MAIN PROGRAM FOR INVERTING THE DF
%  MODEL OF THE DERIVATIVE GAIN GIVEN IN TABLE 5.1
% THE ASSUMED NONLINEARITY GAIN IS DEPICTED IN FIGURE 5.1
%
%   ampl      is the amplitude of the error signal
%   sw          is an arbitrary frequency
%  desgain   is the derivative gain in Table 5.1
%
global ampl sw desgain
% arbitrarily set the frequency sw to 5 rad/sec; since
% the nonlinearity is memoryless any frequency would work
sw=5;
% input the amplitudes
ampl=[0.0628  0.0804  0.1005  0.1608  0.3215  0.6430  1.2811
2.5621];
%input the gains at the defined amplitudes
desgain=[0.0433 0.0447 0.0496 0.0688 0.1292 0.1673 0.1931
0.2062];
%provide an initial guess
x0=[.2 0. 0.1 0.5];
%minimize the objective function (5.5)
x=fminsearch('errorinv',x0);
% generate the describing funcion model of the
% synthesized nonlinearity
[rfit,ifit,amchk,phchk]=getsidfinv2(sw,ampl,10,x);
% for comparison purposes generate the describing
% function model of starting solution
[rfit2,ifit2,amchk2,phchk2]=getsidfinv2(sw,ampl,10,x0);
% plot results
plot(ampl,rfit,'r',ampl,desgain,'b',ampl,rfit2,'c')
%
```

Fig. 5.5 Listing of the software for describing function inversion. Source: (Nassirharand 2009a) © Elsevier

```
%
%
% The following function calculates the objective
% func   tion value
%
function serr=errorinv(x)
global ampl sw desgain
%  x(1) = delta    x(2) = D   x(3) = S1 x(4) = S2
%
%  give a penlty for unreasonable solutions
if( (x(1) < 0 ) | (x(3) <0) )
   serr=10000;
   return
end
%
% obtain the DF model of the candidate
% nonlinear function
%
[real imag am ph]=getsidfinv2(sw,ampl,20,x);
sampl=size(ampl);na=sampl(2);
serr=0;
%
% calculate objective function (5.5)
for i=1:na
   serr=serr+((desgain(i)-real(i))^2);
end
```

Fig. 5.5 (continued)

nonlinear function (i.e., *getsidfinv2*), and the function that calculates the output of the assumed nonlinear function (i.e., *getyinv2*) are given.

In the main program, the gain vs. amplitude data is supplied, and then the *fminsearch* function of the MATLAB® is invoked to minimize Objective Function (5.5). This followed by verification of the results, and the main program ends. The function *errorinv* first imposes any penalty functions that may be associated with the undesirable values for the parameters of the assumed nonlinear function. Then, the describing function model of the assumed nonlinear function is generated followed by evaluation of Objective Function (5.5). Function *getsidfinv2* calculates the describing function model of the nonlinear function; the user must supply the function *getyinv2* that calculates the outputof the nonlinear function that is required

```
%
% The following function generates the describing function
% model of the
% candidate nonlinear function
%
function [gr,gi,amchk,phchk]=getsidfinv2(w,ampl,kmax,x)
global amptd freq
sw=size(w);iw=sw(2);sampl=size(ampl);iampl=sampl(2);
%
% set Fourier integral convergence criteria
%
amepsln=0.05;phsepsln=5;
for iam=1:iampl
%start of the amplitude loop
   amptd=ampl(iam);
   for ifreq=1:iw
% start of the frequency loop
      freq=w(ifreq);
      T=2*pi/freq;
      for k=1:kmax
% start of the Fourier integral loop
         if(k==1)
            t0=0;
         else
            t0=tfinal;
         end
         tfinal=k*T;
%
%calculate the output of the nonlinear function
%
         [t,y]=getyinv2(t0,tfinal,amptd,freq,x);
%
% calculate Fourier integrals
%
         intgd1=y.*cos(freq*t);fr=trapz(t,intgd1);
         intgd2=-y.*sin(freq*t);fi=trapz(t,intgd2);
         hr(k)=fr*freq/(amptd*pi);
         hi(k)=fi*freq/(amptd*pi)
```

Fig. 5.5 (continued)

for evaluation of the Fourier integrals as discussed in Chap. 2. Function *getyinv2* is user-supplied, and it calculates nonlinearity function for the assumed sinusoidal input.

```
 % calculate gain
         am(k)=sqrt(hr(k)^2+hi(k)^2);
% calculate phase
         phi(k)=atan2(hi(k),hr(k));
         if (k==kmax)
% Fourier integrals did not converge; store the latest values
            gr(iam,ifreq)=hr(k);
            gi(iam,ifreq)=hi(k);
         end
         if( k >=2 )
%  calculate the error in gain using the most two recent values
            amchk(iam,ifreq)=abs( (am(k)-am(k-1)) )/ (abs(am(k)));
%  calculate the error in phase using the most two recent values
            phchk(iam,ifreq)=abs(phi(k)-phi(k-1));
%  check convergence of Fourier integrals
            if(amchk(iam,ifreq) <= amepsln)
               if(phchk <= phsepsln)
%  Fourier integrals converged; store the values
                  gr(iam,ifreq)=hr(k);
                  gi(iam,ifreq)=hi(k);
                  break
               end
            end
         end
      end
   end
end
```

Fig. 5.5 (continued)

5.4 Example: Single-Variable Case: Servo System

5.4.1 Nonlinear PID Controller Design

Consider the SISO servo system that was also considered in Sect. 2.2.1.2. It is desired for the closed-loop system to mimic a second-order system with a damping ratio of 0.307 and a natural frequency of 43.22; this approximately corresponds to a settling time of 0.3 s and an overshoot of 37%. These performance measures were originally achieved by Taylor and Strobel (1985) when designing a nonlinear PID controller. For comparison purposes, the same performance measures are selected. Algorithm 5.1 is executed as follows. Steps 1 and 2 are similar to that presented in Sect. 4.4. After execution of steps 3 and 4, the results of step 5 are summarized in Table 5.1 (Nassirharand and Karimi 2006).

```
%
%The following function calculates the output of the
%nonlinearity for a sinusoid input
%
function [t,y] =getyinv2(t0,tfinal,amptd,freq,x)
% m1=S1 and m2=S2 see Figure 5.1
delta=x(1);d=x(2);m1=x(3);m2=x(4);
t=t0:0.01:tfinal;
u=amptd*cos(freq*t);su=size(u);nu=su(2);
%Program the assumed nonlinear function here:
for i=1:nu
   sgnu=sign(u(i));absu=abs(u(i));
   if(absu <delta)
      y(i)=m1*absu;
   elseif (absu == delta)
      y(i)=m1*absu+d;
   else
      y(i)=m1*delta+d+m2*(absu-delta);
   end
   y(i)=sgnu*y(i);
end
```

Fig. 5.5 (continued)

Table 5.1 The discrete controller gains at various operating regimes. Source: (Nassirharand and Karimi 2006) © Elsevier

No.	e_i	K_P	K_I	K_D
1	0.0628	5.97	37.4	0.043
2	0.0804	5.13	36.0	0.045
3	0.1005	4.71	36.0	0.050
4	0.1608	4.38	39.5	0.069
5	0.3215	6.81	55.9	0.129
6	0.6430	8.89	64.0	0.167
7	1.2811	9.84	67.7	0.193
8	2.5621	10.5	68.5	0.206

Table 5.2 The parameters of the synthesized nonlinear function gains. Source: (Nassirharand and Karimi 2006) © Elsevier

	S_1	S_2	D	δ
K_P	5.10	11.16	0.0	0.189
K_I	36.77	72.81	0.0	0.126
K_D	0.047	0.215	0.0	0.132

With reference to Fig. 5.1, the results of describing function inversion (step 6) are summarized in Table 5.2 (Nassirharand and Karimi 2006).

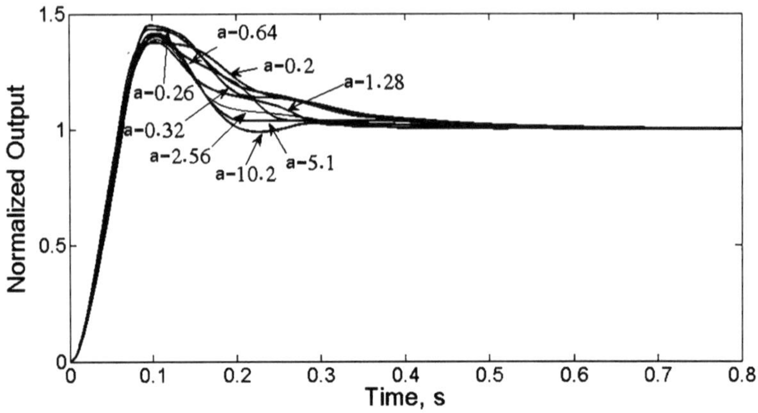

Fig. 5.6 The normalized step responses with various amplitudes of excitation signal

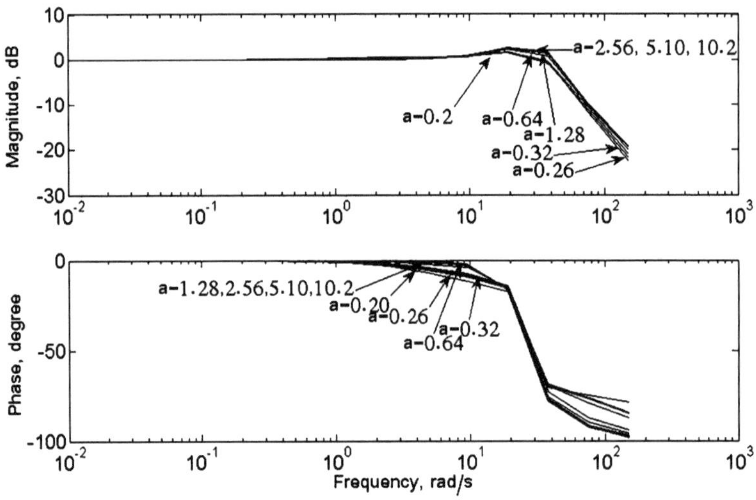

Fig. 5.7 Stability test for the nonlinear PID design

Finally, the design is verified, and the results of the last step are shown in Figs. 5.6 and 5.7.

As was mentioned before, Eqs. 5.2 and 5.3 may be evaluated for frequencies between 0 and 150 rad/s to measure the percent reduction in sensitivity; in this case, $\sigma_0 = 1924$, $\sigma_1 = 221$, and $R = 88.5$. In other words, 88.5% reduction in sensitivity is achieved.

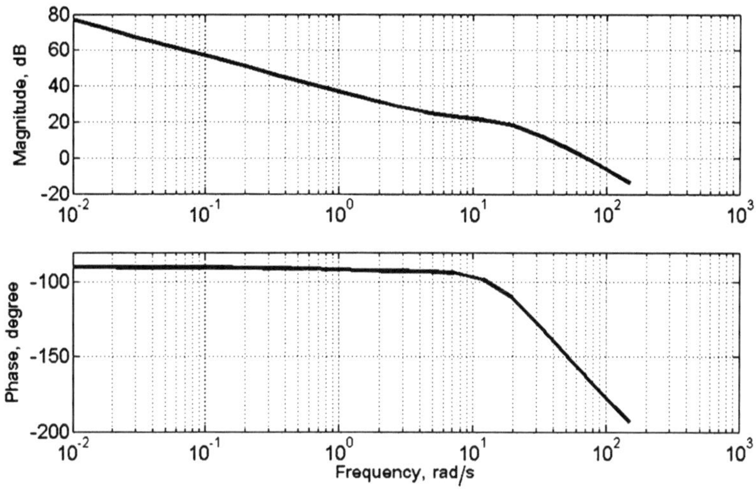

Fig. 5.8 The SIDF model of the open-loop system comprised of the nominal compensator, C^*, and the nonlinear plant

5.4.2 *Nonlinear Lead-Lag Controller Design (Nassirharand and Mousavi Firdeh 2008)*

The problem statement is the same as that given in Sect. 5.4.1 with the difference that a nonlinear lead-lag controller is required to be designed. Steps 1 and 2 are identical to those described in Sect. 4.4. In step 3, the middle model (at excitation of 0.325) is selected as the nominal model; a linear model is identified using *invfreqs* function of MATLAB®, and the MATLAB® function described in Sect. A.1.1 is used to design a lead compensator. The compensator gain is adjusted to obtain a satisfactory step response when the resulting nonlinear system is simulated. The designed lead compensator is of the following form:

$$C^* = 1.5\,(0.3514s + 3.36)\,/\,(0.03303s + 1) \qquad (5.6^*)$$

In step 4, the desired open-loop behavior is determined by obtaining the pseudofrequency response of the open-loop system comprised of the nominal compensator, C^*, and the nonlinear plant. The desired open-loop system behavior is shown in Fig. 5.8.

This quasi-linear model serves as the target open-loop pseudofrequency response model at all operating regimes; this target model is used in optimization in order to obtain the parameters of the lead compensator at various operating regimes that cause insensitivity of the nonlinear open-loop system to various amplitude levels of

*Source: (Nassirharand and Mousav Firdeh 2008) © Springer

Table 5.3 Parameters of the synthesized nonlinear functions. Source: (Nassirharand and Mousav Firdeh 2008) © Springer

	δ_1	δ_2	d_1	d_2
A_0	0.1486	0.2811	0.1822	0.1452
A_1	0.1762	0.5761	0.0079	0.1111
B_1	0.0583	0.2373	0.0006	−0.0001
	d_3	m_1	m_2	m_3
A_0	0.3085	1.4951	1.1223	6.9811
A_1	0.3338	0.4475	0.4105	1.2591
B_1	0.0005	0.0282	0.0168	0.0286

excitation. The target model has approximately a 30-degree phase margin and an approximate gain margin of 3.

In step 5, the set of linear lead compensators at various excitation amplitudes is obtained by minimizing Objective Function (5.1). The controller parameters are: A_0 = {6.1141, 5.0390, 4.3959, 3.3293, 3.3834, 5.0765, 5.9311, 6.3986, 6.7062, 6.8369}, A_1 ={0.6507, 0.6001, 0.5737, 0.5447, 0.7374, 1.0177, 1.1620, 1.2460, 1.3322, 1.1587}, and B_1 ={0.0423, 0.0380, 0.0344, 0.0241, 0.0235, 0.0259, 0.0271, 0.0272, 0.0264, 0.0302}. Where, the parameters of the lead compensator are defined as below:

$$C = (A_1 s + A_0) / (B_1 s + 1) \tag{5.7*}$$

The constants are nonlinear function of the error signal: (a) for a given value of the error signal, the nonlinearities (obtained in step 5) are evaluated and this gives a set of constants; (b) in SIMULINK® implementation, the values obtained in (a) are divided by error value; and (c) the values obtained in (b) sit in place of the lead and/or lag coefficients and the resulting controller is excited by the error signal.

In step 6, the parameters of the synthesized nonlinear functions (see Fig. 5.2) for compensator parameters are defined in Table 5.3.

The quality of fit is shown in Figs. 5.9, 5.10, and 5.11. These figures show the results of describing function (DF) inversion process for A_0, A_1, and B_1. The nonlinearity parameters are optimized in a fashion that DF model of that nonlinearity mimics the desired amplitude-dependent gains of the compensator parameters.

Finally, in step 7 the design is verified using numerical simulation. The normalized step-response plots of the closed-loop system at various operating regimes of interest are depicted in Fig. 5.12. This figure demonstrates the results of design verification step. The normalized step responses at various amplitudes of reference signals indicate that closed-loop system behavior is fairly insensitive to the amplitude level of the command signal.

For stability, the describing function models of the synthesized nonlinear closed-loop system at various operating regimes are generated. The stability test is shown

*Source: (Nassirharand and Mousav Firdeh 2008) © Springer

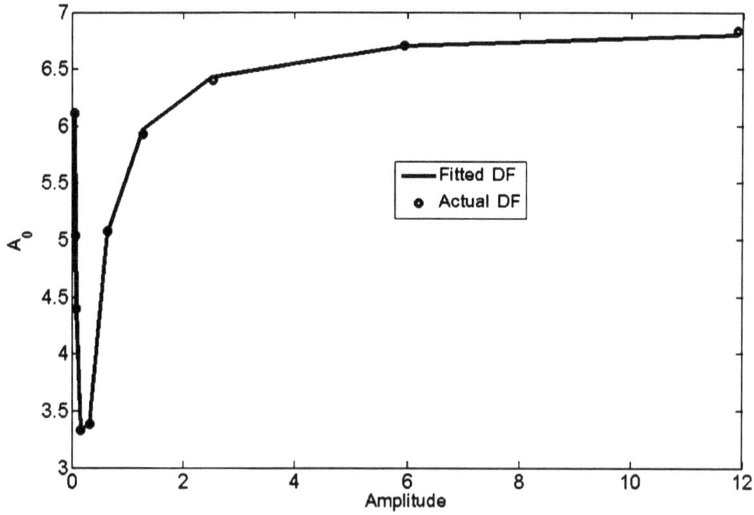

Fig. 5.9 The results of describing function inversion process for A_0. Source: (Nassirharand and Mousav Firdeh 2008) © Springer

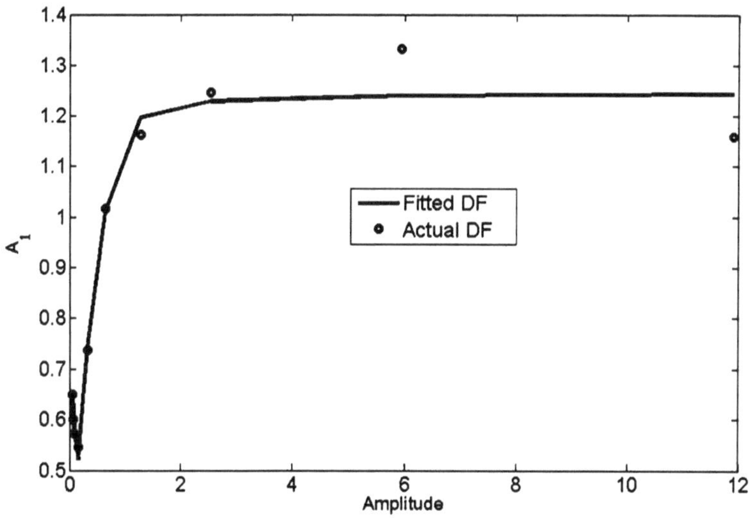

Fig. 5.10 The results of describing function inversion process for A_1. Source: (Nassirharand and Mousav Firdeh 2008) © Springer

in Fig. 5.13. Also, it is apparent that closed-loop system is fairly insensitive to the amplitudes of excitations. As was mentioned in step 7 of Algorithm 5.1, σ_0 and σ_1 may be evaluated to measure the percent reduction in sensitivity; in this case, $\sigma_0 = 2308$, $\sigma_1 = 322$, and $R = 86$. In other words, 86% reduction in sensitivity is achieved.

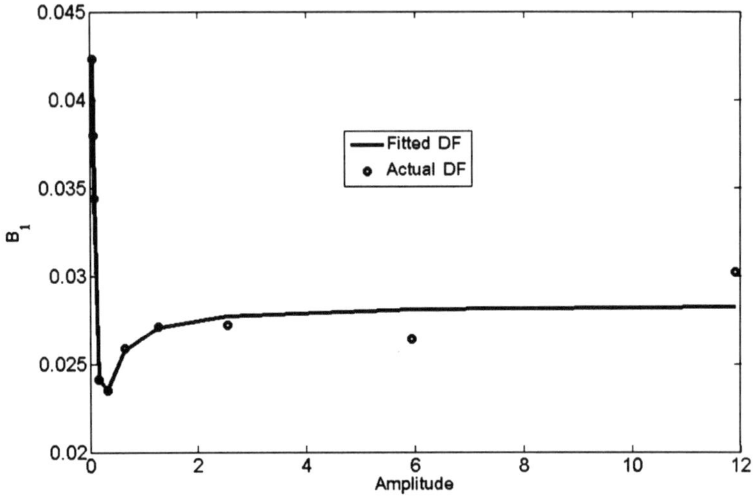

Fig. 5.11 The results of describing function inversion process for B_1. Source: (Nassirharand and Mousav Firdeh 2008) © Springer

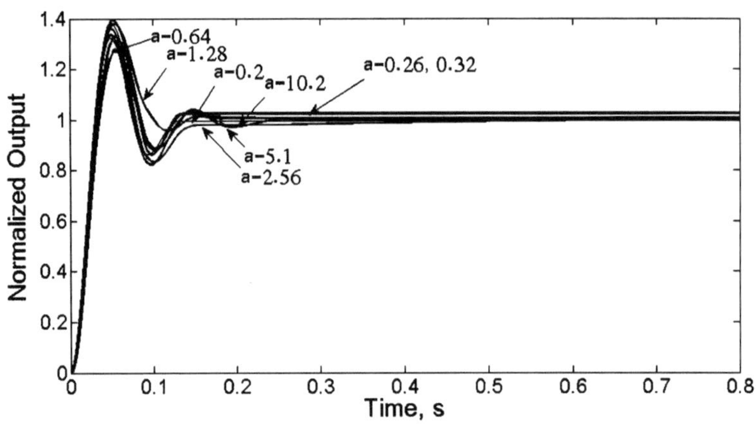

Fig. 5.12 Normalized step-response plots of the synthesized nonlinear control system. Source: (Nassirharand and Mousav Firdeh 2008) © Springer

5.4.3 Nonlinear H_∞ Controller Design

The problem statement is the same as that given in Sect. 5.4.1 with the difference that a nonlinear H_∞ controller is required to be designed. The structure of the nonlinear H_∞ controller is shown in Fig. 5.14 (Nassirharand and Mousavi Firdeh 2009c). Steps 1 and 2 are identical to those described in Sect. 5.3.1. In step 3, the model with the highest gain amplification effect (at excitation of 0.8) is

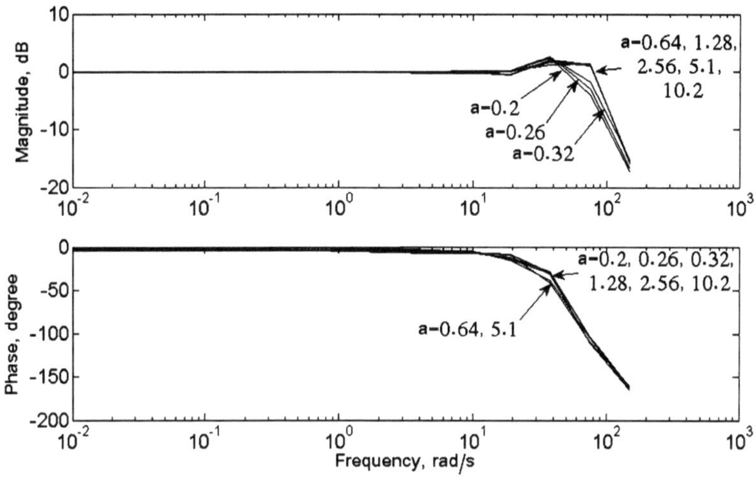

Fig. 5.13 Stability test for the nonlinear lead-lag controller. Source: (Nassirharand and Mousav Firdeh 2008) © Springer

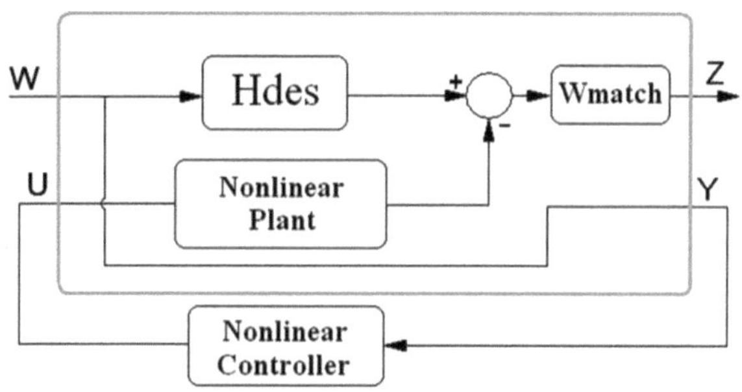

Fig. 5.14 The considered nonlinear fractional transformation. Source: (Nassirharand and Mousavi Firdeh 2009c) © Wiley

selected as the nominal model; a linear model is identified using *invfreqs* function of MATLAB®, and the MATLAB® function *hinfsyn* is used to design a H_∞ controller. The controller gain is adjusted to obtain a satisfactory step response when the resulting nonlinear system is simulated. The designed H_∞ controller is of the following form:

$$C^* = 5.0\frac{1447s + 2.154e4}{270.3s + 1.019e4} \tag{5.8}$$

In step 4, the desired open-loop behavior is determined by obtaining the open-loop sinusoidal-input describing function model comprised of $C^*$and the nonlinear plant. In step 5, the set of linear H_∞controllers at various excitation amplitudes are obtained as outlined previously; those are given below:

Table 5.4 Parameters of the identified nonlinear functions

	δ_1	δ_2	δ_3	d_1	d_2
A_0	0.0102	0.0229	2.1452	−0.0029	−0.0685
A_1	0.0011	0.0477	0.0599	−0.2709	0.3943
B_0	0.0120	0.0679	2.1806	0.0047	0.0001
B_1	0.0327	0.0328	0.9938	0.0097	0.0233
	m_1	m_2	m_3	m_4	
A_0	4.3368	−2.6090	1.8661	5.0720	
A_1	0.0025	7.4417	−0.2044	9.0010	
B_0	0.3462	1.1701	0.7885	1.8399	
B_1	3.8129	14.9105	1.1075	0.6282	

$$A_0 = 1e5 \begin{Bmatrix} 1.0537, 0.8967, 1.0580, 1.1145, 1.4041, 1.6809, 1.9423, 1.8791, \\ 2.2523, 3.5026 \end{Bmatrix}$$

$$A_1 = 1e3 \begin{Bmatrix} 7.3540, 6.7826, 6.8800, 7.0189, 7.9919, 8.7454, 9.1093, 8.7004, \\ 8.6104, 8.9454 \end{Bmatrix}$$

$$B_0 = 1e4 \begin{Bmatrix} 0.6893, 0.6927, 0.8584, 1.0524, 0.8955, 0.8370, 0.8690, 0.8058, \\ 0.9209, 1.3263 \end{Bmatrix}$$

$$B_1 = \begin{Bmatrix} 383.8594, 367.8280, 385.5476, 266.7148, 172.5880, 155.4505, \\ 138.2746, 121.1289, 80.5201, 80.5201 \end{Bmatrix}$$

Where the parameters of the H_∞ controller are defined as below:

$$C(s) = \frac{A_1 s + A_0}{B_1 s + B_0} \tag{5.9}$$

In step 6, controller parameters are inverted. With reference to Fig. 5.3, the parameters of the synthesized nonlinear functions for controller parameters are defined in Table 5.4. The quality of fit is shown in Figs. 5.15, 5.16, 5.17, and 5.18.

Finally, in step 7, the design is verified using numerical simulation. For stability test, see the sinusoidal-input describing function models of the final synthesized closed-loop system as depicted in Fig. 5.19. Also, the normalized step-response plots of the closed-loop system at various operating regimes of interest are depicted in Fig. 5.20. From examination of this figure, it may be concluded that amplitude insensitivity objective is satisfied. As was mentioned before, Eqs. 5.2–5.4 may be evaluated to measure the percent reduction in sensitivity; in this case, $\sigma_0 \approx 2308$, $\sigma_1 \approx 350$, $\sigma_1 \approx 477$, $R_1 \approx 85$, and $R_2 \approx 79$. In other words, 85% reduction in sensitivity is achieved when the nonlinear controller is used, and 79% reduction in sensitivity is achieved with the nominal linear controller, C^*, utilized.

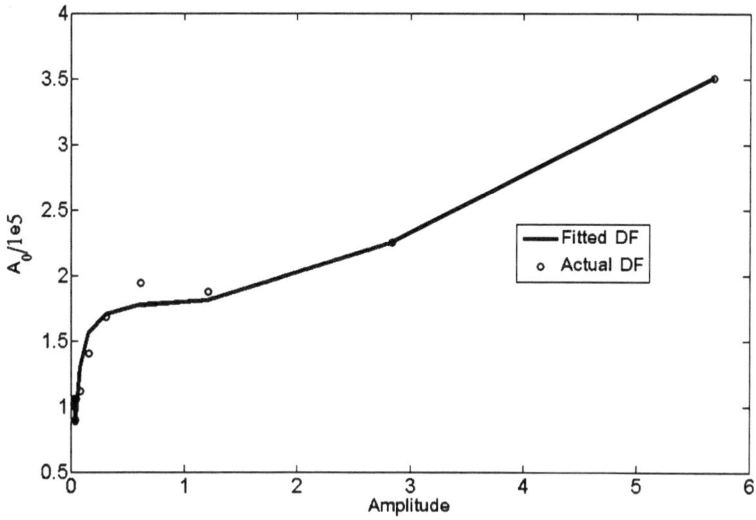

Fig. 5.15 Results of describing function (*DF*) inversion for controller parameter A_0

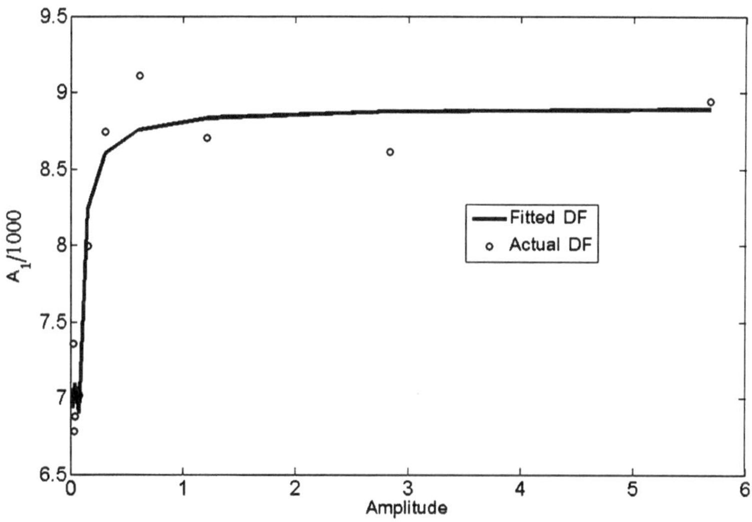

Fig. 5.16 Results of describing function (*DF*) inversion for controller parameter A_1

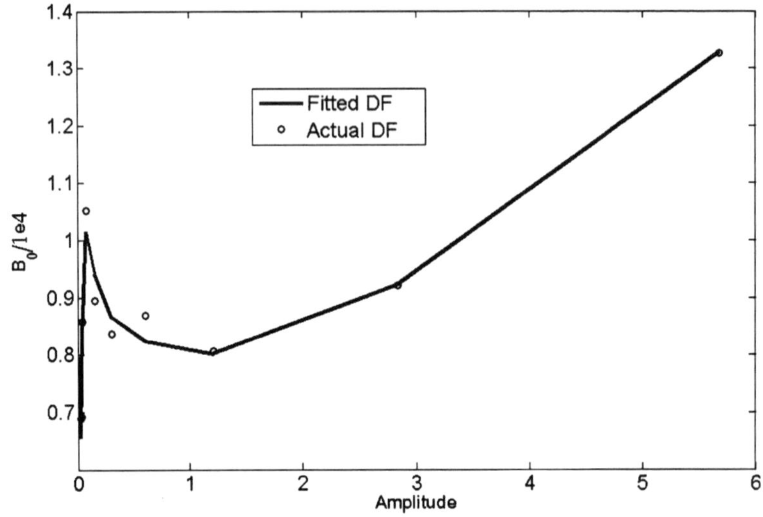

Fig. 5.17 Results of describing function (*DF*) inversion for controller parameter B_0

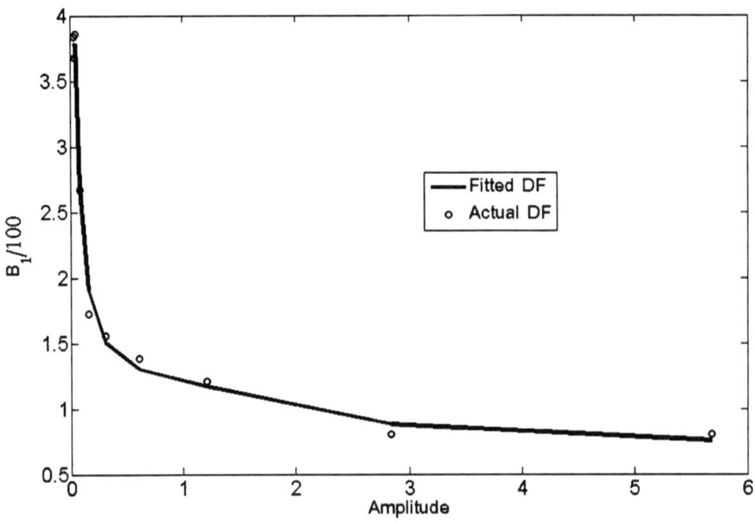

Fig. 5.18 Results of describing function (*DF*) inversion for controller parameter B_1

5.5 Example: Multivariable Case: Bank Angle Control of a UAV

The UAV that is studied in Sect. 4.5 is considered (see Fig. 4.3). It is desired to design a robust controller for this UAV. As was mentioned in Sect. 4.5, the effectiveness of a single-range linear controller is investigated by Nassirharand

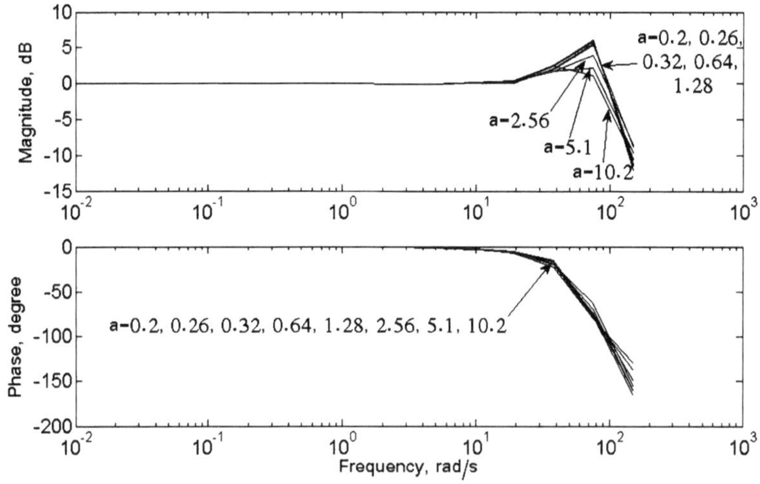

Fig. 5.19 Stability test of the closed-loop system with the nonlinear H_∞ controller

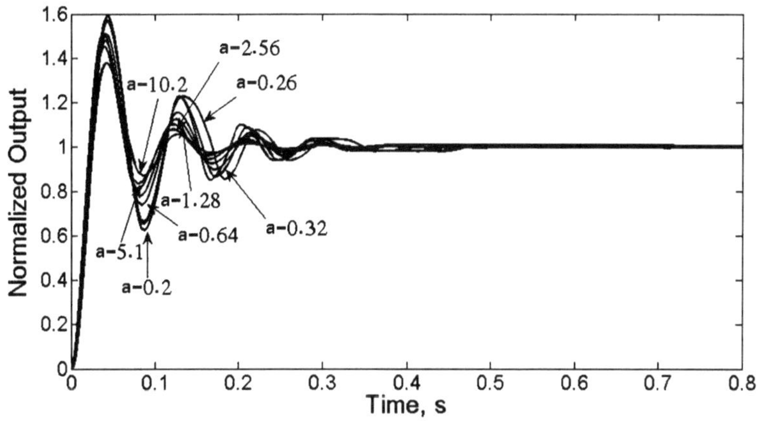

Fig. 5.20 Normalized step responses of the designed nonlinear closed-loop system

and Mousavi Firdeh (2009b). In this case, a linear H_∞ controller (based on one operating regime) is also designed; the corresponding step-response plots are shown in Fig. 4.10. It is apparent that this is not a satisfactory design. The nonlinear controller synthesis procedure is executed as follows.

In step 1, the desired linear model has a settling time of about 2 s with no overshoot. In step 2, the SIDF models for the amplitude set $a = \{0.75, 1, 2, 4, 8, 16, 32\}$ and a frequency range from 0.1 to 1,000 rad/s are generated as outlined in the previous section. The results are depicted in Fig. 4.4. In step 3, arbitrarily, the model at $a = 1$ is selected as the nominal model, and a linear proportional controller is designed. The multivariable gains are $K_\mathrm{p} = [-4384, -1912]$. In

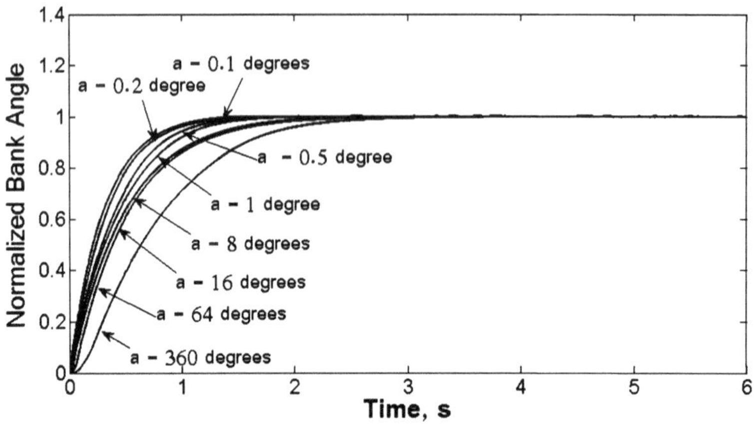

Fig. 5.21 Normalized step-response plots of the closed-loop system with a nonlinear controller and the nonlinear UAV. Source: (Teh 2010) © S.H. Teh

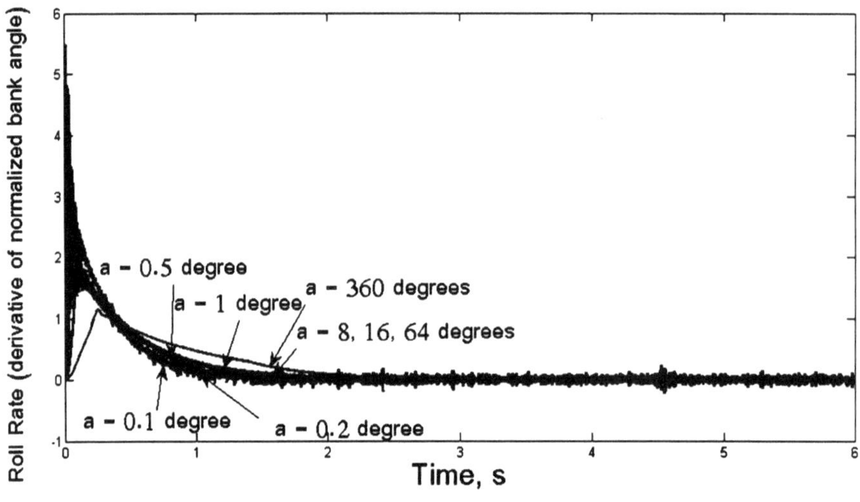

Fig. 5.22 The normalized roll-rate plots of the closed-loop system with a nonlinear controller and the nonlinear UAV. Source: (Teh 2010) © S.H. Teh

order to characterize the controller gains at small, medium, and large signals, the following amplitudes are selected: $E = \{0.05, 0.1, 0.3, 0.5, 0.7, 1, 2\}$. In step 4, with $E = [0.7, 0.7]$ and the frequency range from 0.1 to 50 rad/s, the desired open-loop DF model is generated. In step 5, the *fminsearch* function of the MATLAB® that uses a Simplex search technique (Reklaitis et al. 1983) is used, and the Objective Function (5.5) at various operating regimes is minimized; the data for frequencies below 4 rad/s are weighted ten times more than the others to improve on the steady-state conditions. The following controller gains are obtained:

$$K_{\mathrm{p}} = 1000 \begin{bmatrix} \{-5.26, -5.33, -4.84, -4.69, -4.44, -4.40, -4.38\} \\ \{-1.51, -1.55, -1.68, -1.79, -1.94, -2.06, -2.12\} \end{bmatrix}$$

In step 6, the SIDF models of the controller gains obtained in the previous step are inverted, and nonlinear functions of the type shown in Fig. 5.2 are identified. The results of inversion follow:

$$\delta_1 = [0.1349, 1.19e5]\,, \quad \delta_2 = [51.08, 3.63e6]\,,$$

$$d_1 = [0.0, -20.53]\,, \quad d_2 = [0.0, 74.77]\,,$$

$$d_3 = [0.0, 1.72e5]\,, \quad m_1 = [5.35e3, 1.95e3]\,,$$

$$m_2 = [4.36e3, -1.18e6]\,, \quad m_3 = [2.48e3, -4.06e6]\,.$$

In step 7, the normalized step responses of the closed-loop system with the synthesized nonlinear multivariable controller are depicted in Figs. 5.21 and 5.22. The settling time is about 1.5 s with no overshoot. A fourth-order nonlinear H_∞ controller is designed by Colgren (2004); in that design a settling time of 40 s with no overshoot is achieved (see Fig. 4.12).

Chapter 6
Experimental Study: Unstable SISO Systems

6.1 Background

The presented methods developed so far are limited to stable nonlinear systems. This is mainly due to the fact that Fourier integrals are not able to converge if the outputs are not bounded; hence, the corresponding describing function models are not obtainable. In this chapter, an approach is outlined that would allow the techniques developed so far to be applied to unstable systems (Li and Nassirharand 2011). An experimental rig is designed and fabricated to demonstrate the approach.

6.2 Experimental Rig

The experimental rig is shown in Figs. 6.1, 6.2, and 6.3.

This setup is comprised of an AC motor with a rated output of 2 KW, rated torque of 19.1 Nm, maximum instantaneous torque of 57.3 Nm, rated speed of 1,000 rpm, maximum speed of 2,000 rpm, rotor inertia of 98.4e-4 kg·m^2, rated power rate of 37.0 KW/s, and a total weight of 27.4 kg. The encoder has a shaft diameter of 8 mm with a resolution of 2,500 pulses/revolution, and it requires a power supply of 5VDC. The gear ratio is 5:1. An additional pulley with mass configuration is added to increase the degree of instability of the plant by constantly pulling down on the pendulum from its equilibrium position as shown in Fig. 6.4.

The pendulum is an aluminum bar with an outer diameter of 22 mm, an inner diameter of 15 mm, and a length of 250 mm. The electronic instrumentation corresponds to National Instruments products with LabView® software. There are also mechanisms for adjustable Coulomb friction and adjustable backlash that are not utilized in this research.

A. Nassirharand, *Computer-aided Nonlinear Control System Design: Using Describing Function Models*, DOI 10.1007/978-1-4471-2149-7_6,
© Springer-Verlag London Limited 2012

Fig. 6.1 The overall picture of the experimental rig. Source: (Li and Nassirharand 2011) © Sage

6.3 Nonlinear PID Synthesis

The problem statement is to develop an approach for synthesis of nonlinear PID controllers for highly nonlinear unstable systems with no restrictions on the number of nonlinear terms, nonlinearity type (could be continuous and/or discontinuous), arrangement (could be in forward path and/or in the feedback path), and system order; the approach is to be demonstrated by applying the procedure to a highly nonlinear inverted pendulum experimental setup.

The developed synthesis procedure is composed of eight primary steps as follows (Li and Nassirharand 2011):

1. Stabilize the unstable nonlinear system around the operating regimes of interest. For problems that a mathematical model is available, one could use a variety of techniques such as sliding mode techniques (Utkin et al. 1999). For cases that a mathematical model is not available, intuition may be used. For example,

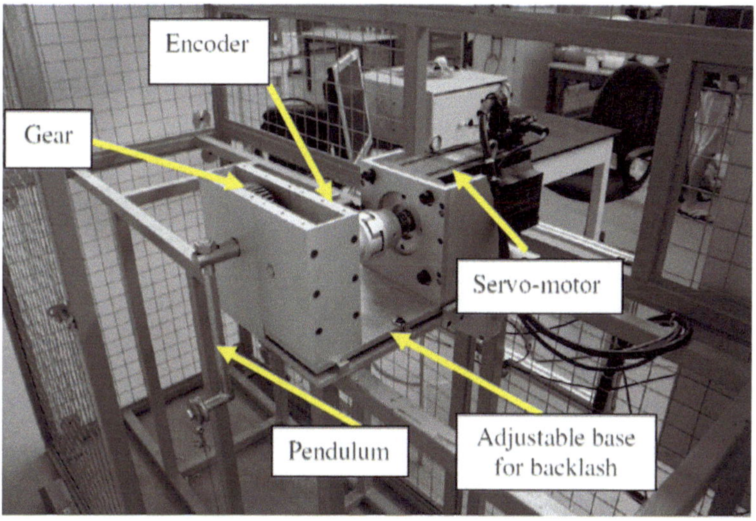

Fig. 6.2 Servomotor and pendulum assembly. Source: (Li and Nassirharand 2011) © Sage

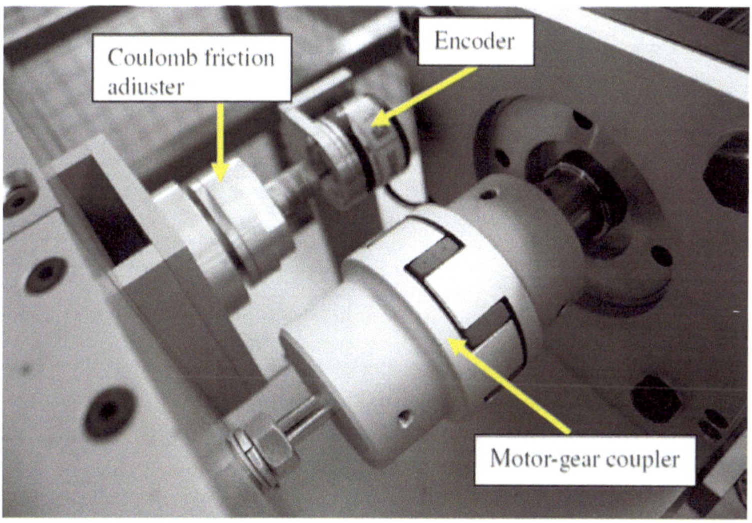

Fig. 6.3 Coulomb friction adjuster and encoder. Source: (Li and Nassirharand 2011) © Sage

for the experimental setup considered here, it is found that a linear proportional controller with a gain of ten may be used to stabilize the system.

2. Generate the sinusoidal-input describing function models of the stabilized closed-loop system around the operating regimes of interest as outlined in Chap. 2. The user may choose to fit a linear model to each of the identified closed-loop frequency domain models. For example, for the example problem

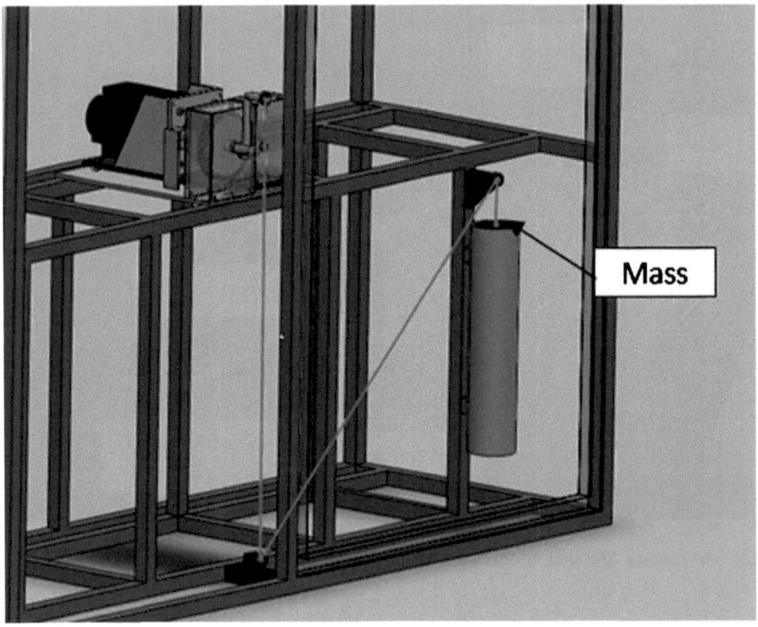

Fig. 6.4 Pulley system to increase the degree of instability. Source: (Li and Nassirharand 2011) © Sage

under study at nominal conditions (i.e., a reference amplitude of 20 degrees), the following closed-loop transfer function may be identified by using the *invfreqs* function of MATLAB®.

$$G_{\mathrm{cl}} = \frac{2.055s + 7.065}{s^2 + 3.304s + 6.385} \tag{6.1}$$

With the known stabilizing controller of step 1, and the closed-loop frequency domain model of the previous step, use algebra and determine the corresponding unstable open-loop frequency domain models denoted $G(j\omega; a_i)$. For the inverted pendulum problem, the corresponding unstable open-loop frequency domain model at the nominal condition is determined to be as follows:

$$G_{\mathrm{ol}} = \frac{0.2055s + 0.7065}{s^2 + 1.249s - 0.6803} \tag{6.2}$$

The set of unstable open-loop models are determined using algebra, and they are depicted in Fig. 6.5. The spread in the pseudo-Bode plots indicate that system is highly nonlinear. If this were a linear plant, then there would be only one pseudo-Bode plot for all excitation amplitudes.

3. Select one of the models of the previous step as the nominal model, and design a PID controller. Then use the MATLAB® function developed in Appendix A

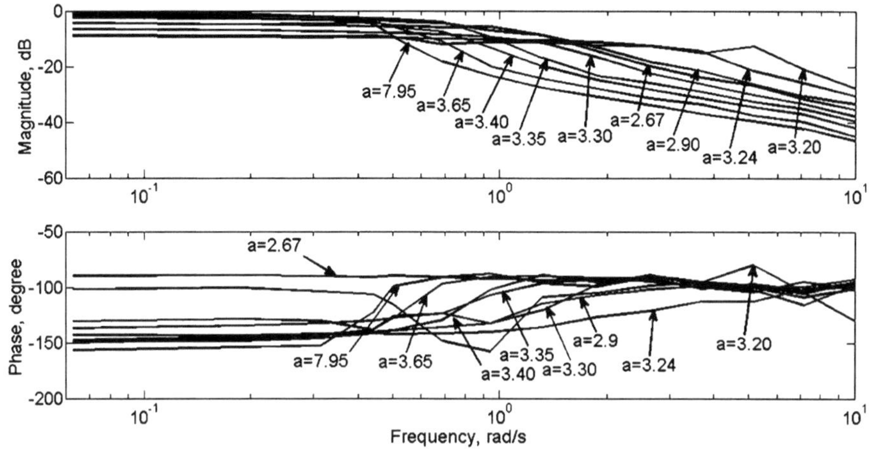

Fig. 6.5 Frequency domain models of the unstable plant (*a* corresponds to the amplitude of the sinusoid at the input to the plant). Source: (Li and Nassirharand 2011) © Sage

or Appendix B, and design a linear PID controller denoted C^*. In this case, the model corresponding to Eq. 6.2 is selected, and the MATLAB® function *cad_controller* developed in Appendix B is used, and the following PD (integral gain turns out to be negligible, i.e., $K_I \approx 0$) controller is designed.

$$C^*(s) = 0.09272s + 20.01 \tag{6.3}$$

The primary inputs to the *cad_controller* function are the plant given by Eq. 6.2 and a desired closed-loop transfer function; in this case, a second-order transfer function with a damping ratio of 0.9 and a settling time of 0.5 s is specified to be the desired closed-loop transfer function. The output of the above-mentioned function is the PD controller given by Eq. 6.3.

4. Generate the closed-loop sinusoidal-input describing function model of the system comprised of the controller of the previous step and the nonlinear plant similar to what is done in step 2. Then, with the known desired closed-loop SIDF model and the known controller of step 3, use algebra, and determine the desired open-loop frequency domain model denoted C^*G^*. For the inverted pendulum problem, this is shown in Fig. 6.6.
5. Design a set of linear PID controllers at various operating regimes of interest that mimic the desired open-loop behavior obtained in the previous step. This is accomplished by minimizing the following objective function:

$$E(j\omega) = |C^*G^* - C_i(j\omega)\, G(j\omega, e_i\, |C_i(j\omega)|)| \tag{6.4}$$

$G(j\omega, e_i\, |C_i(j\omega)|)$ is evaluated by using the $G(j\omega, a_i)$ data of step 1 coupled with interpolation. This will allow one to execute this step with standard available personal computers. The output of this step is a set of values for the

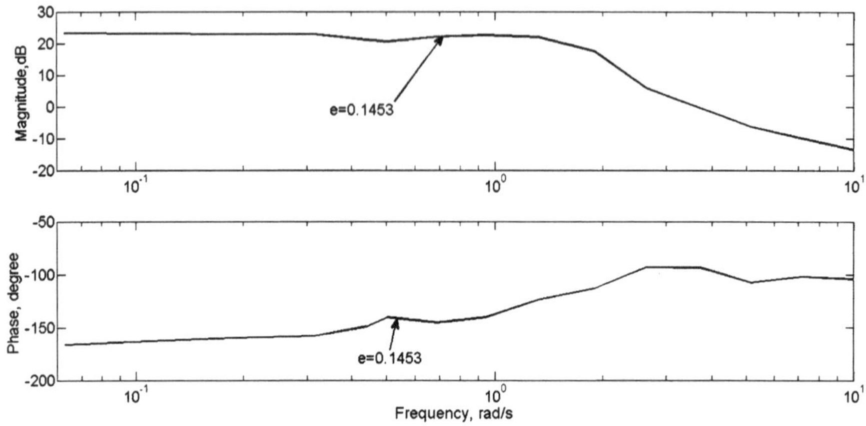

Fig. 6.6 The desired open-loop frequency domain model (*e* is the amplitude of the sinusoid at the input to the controller). Source: (Li and Nassirharand 2011) © Sage

Table 6.1 The amplitude-dependent gains of the PD controller. Source: (Li and Nassirharand 2011) © Sage

e, error	K_P	K_D
0.1338	21.62308	−0.08978
0.1453	19.95128	−0.08963
0.1604	20.48715	0.028779
0.1624	20.23243	0.027646
0.1654	19.86186	0.026103
0.1679	19.56322	0.024987
0.1704	19.27333	0.024041
0.1829	15.85274	0.100826

parameters of the PID compensator at various operating regimes of interest. For the inverted pendulum problem, the optimized values of the PD controller (the gain for integral action is determined to be negligible) are given in Table 6.1. These amplitude-dependent parameters are treated as describing function models of the desired nonlinear gains of the nonlinear controller.

The corresponding open-loop system comprised of the controllers given in Table 6.1 and the unstable plants derived in step 2 are depicted in Fig. 6.7.

6. Determine a set of nonlinear functions whose SIDF models mimic the SIDF models of the compensator parameters that were determined in the previous step. A typical nonlinear function that a user may select is shown in Fig. 5.4. The parameters of this function are determined in such a manner that the SIDF model approximates the amplitude-dependent gains of the compensator parameters at various operating regimes of interest. Such a function will allow for high gains at low amplitudes, constant gains at specific amplitude ranges, and gain reductions or jumps at specific amplitudes. Note that this step is the inverse of activities of step 2. In step 2, the model is available and the SIDF models are to be obtained;

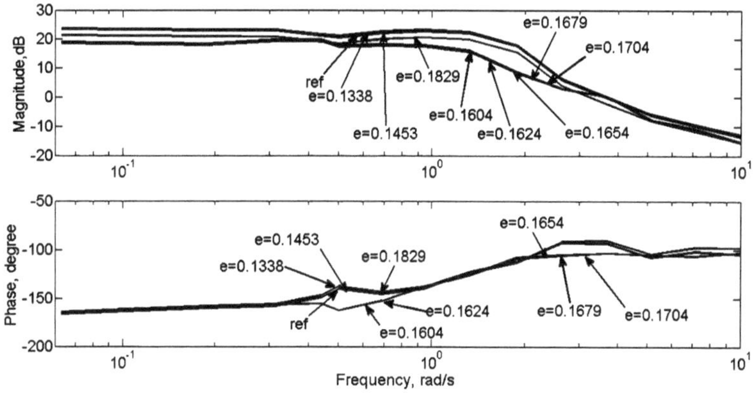

Fig. 6.7 The frequency domain models of the open-loop system comprised of the amplitude-dependent controllers and the unstable nonlinear plant. Source: (Li and Nassirharand 2011) © Sage

Table 6.2 Results of SIDF inversion. Source: (Li and Nassirharand 2011) © Sage

	δ_1	δ_2	d_1	d_2	d_3
K_P	2.14	2.15	2.19	0.444	−6.55
K_D	6.72	9.27	−0.077	0.621	4.98
	m_1	m_2	m_3		
K_P	2.32	2,640.3	−15.83		
K_D	0.621	107.2	−1.938		

in this step, the SIDF models are known, and the nonlinearity model must be obtained. In order to determine the parameters of the nonlinearity function shown in Fig. 5.2, optimization is used, and the *fminsearch* function of the MATLAB® software is utilized. The details of this inversion process are (see Chap. 5): (a) use the describing function generator MATLAB® function from Chap. 2 and generate the describing function model of the nonlinearity, (b) if satisfactory results are obtained, then stop; otherwise, use *fminsearch* function of the MATLAB® in order to modify the independent variables and go to (a).

With reference to Fig. 5.2, for the inverted pendulum example, the results of SIDF inversion using the software outlined in Chap. 5 are given by Table 6.2. The quality of fit is shown in Figs. 6.8 and 6.9.

Finally, the design is verified using experiments. The normalized step-response plots of the closed-loop system at various operating regimes of interest are depicted in Fig. 6.10. This figure demonstrates the results of design verification step. The normalized step responses at various amplitudes of reference signals indicate that closed-loop system behavior is fairly insensitive to the amplitude level of the command signal. It is observed that at lower step magnitudes, there exist high-frequency oscillations. The main cause of this is because at the connection point of the pendulum to the shaft of the motor, there is a two degrees free play. The stability

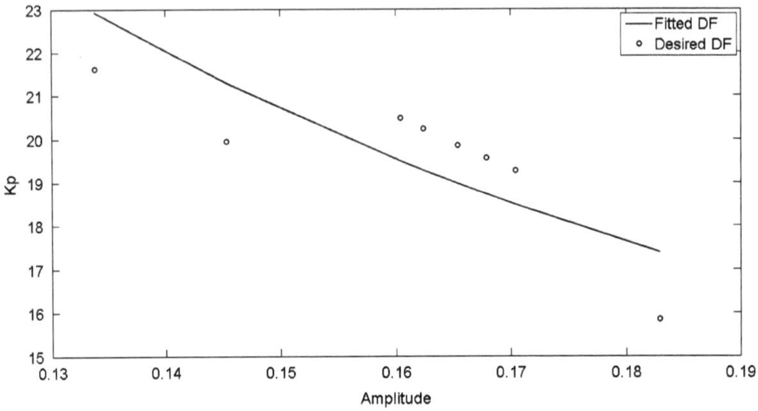

Fig. 6.8 The quality of fit after SIDF inversion for the proportional nonlinear gain. Source: (Li and Nassirharand 2011) © Sage

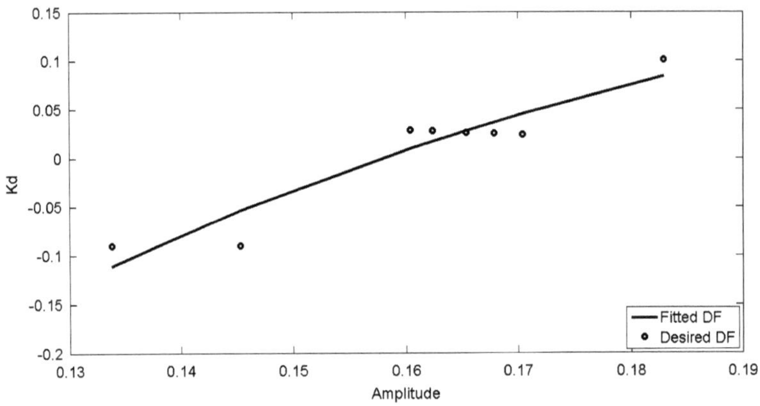

Fig. 6.9 The quality of fit after SIDF inversion for the derivative nonlinear gain. Source: (Li and Nassirharand 2011) © Sage

test is shown in Fig. 6.11. In this figure, the pseudofrequency response plots of the closed-loop system comprised of the designed nonlinear PD compensator and the nonlinear process are successfully generated; hence, stability at operating regimes of interest is demonstrated.

A linear controller based on one frequency domain model is designed for comparison purposes; the corresponding normalized step-response plots are shown in Fig. 6.12.

It is apparent that the single-range linear controller design is not satisfactory (because of sensitivity and large steady-state errors), and the nonlinear PD controller is preferred. Another linear controller based on two frequency domain models is also designed (see Chap. 4); the corresponding normalized step-response plots are shown in Fig. 6.13.

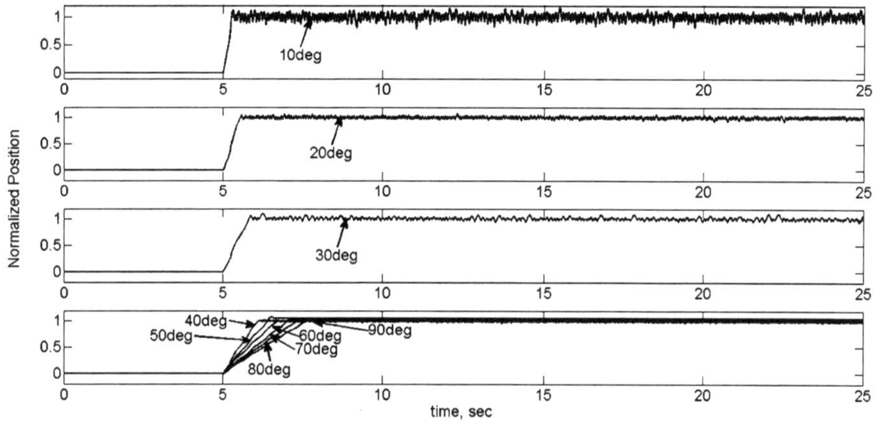

Fig. 6.10 Normalized step-response plots of the synthesized nonlinear closed-loop system. Source: (Li and Nassirharand 2011) © Sage

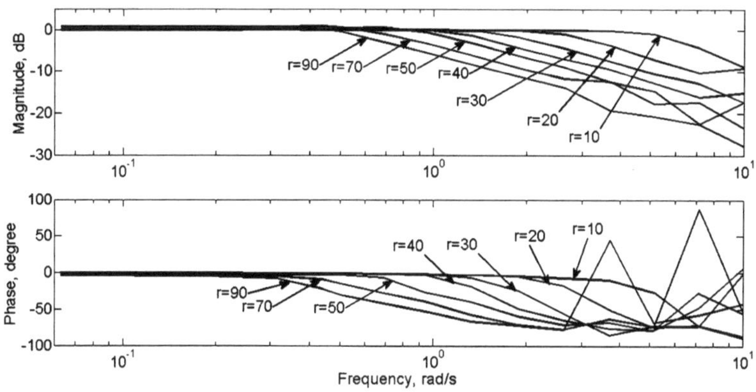

Fig. 6.11 Stability test (*r* corresponds to the amplitude of the sinusoid for the reference signal). Source: (Li and Nassirharand 2011) © Sage

These results are competitive with the nonlinear PD controller results; these results are consistent with prior theoretical work in previous chapters.

6.4 Conclusions

In short, the synthesis procedure involves stabilization of the nonlinear plant followed by determination of the SIDF models of the stabilized closed-loop system. Algebra is used, and the unstable frequency domain open-loop models are obtained followed by a linear controller design at nominal conditions. The desired open-loop

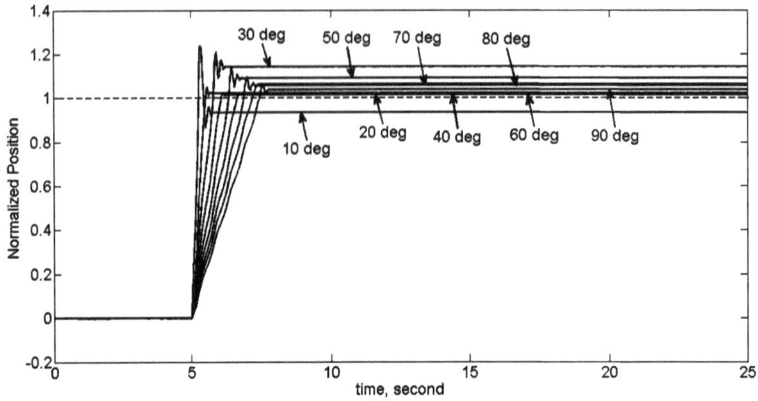

Fig. 6.12 The normalized step-response plots with the single-range linear controller. Source: (Li and Nassirharand 2011) © Sage

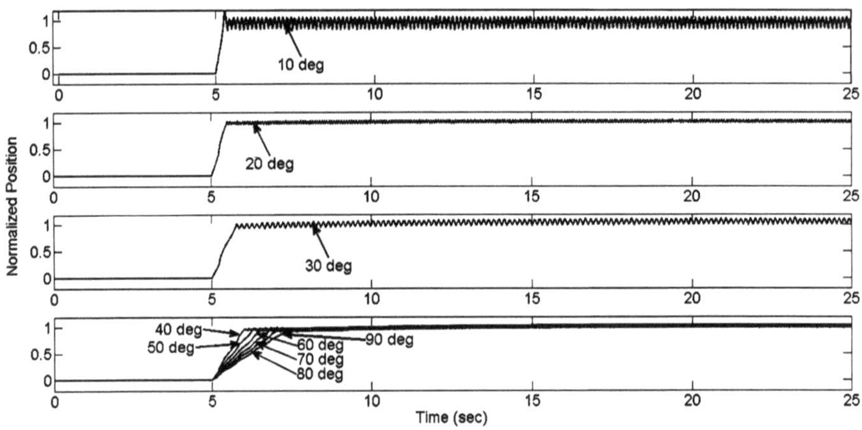

Fig. 6.13 The normalized step-response plots with the dual-range linear controller. Source: (Li and Nassirharand 2011) © Sage

frequency domain model is determined, and a set of linear controllers that force the open-loop frequency domain models at all operating regimes that mimic that of the desired is determined by optimization. The amplitude-dependent gains are inverted to obtain the nonlinear functions describing the nonlinear gains of the controller. Finally, design is verified by obtaining the normalized step-response plots as well as the corresponding closed-loop frequency domain models for stability test. An experimental setup consisting of an AC servomotor and an inverted pendulum is used to demonstrate the procedure. Based on the results of the experimental work, the following conclusions are made: (1) the nonlinear PID synthesis procedure is effective and viable to be applied to highly nonlinear SISO systems, and (2) a stable robust design may be obtained.

Appendix A
Closed-Form Solution to Linear Classical Controllers

A.1 Linear Lead-Lag Controller Design

Consider the linear controller design problem as depicted in Fig. A.1.

Closed-form solution for the design of lead-lag controllers based on an exact model-matching criterion is developed (Nassirharand and Karimi 2004c). The problem statement is to design a linear lead-lag controller for a linear plant in a unity feedback configuration that would satisfy a set of user-defined performance measures. The linear plant is expressed as follows:

$$G_p(j\omega) = \frac{g_1 + g_2 j}{g_3 + g_4 j} \tag{A.1*}$$

The lead-lag controller model is given below:

$$G_c(s) = \frac{a_0 + a_1 s + a_2 s^2}{1 + b_1 s + b_2 s^2}, s = j\omega \tag{A.2*}$$

The desired closed-loop system behavior, or the desired reference linear model, is denoted F:

$$F(j\omega) = \frac{h_1 + h_2 j}{h_3 + h_4 j} = \frac{Y_1}{Y_2} \tag{A.3*}$$

The designed closed-loop system is denoted G_{cl}:

$$G_{cl} = \frac{G_c G_p}{1 + G_c G_p} = \frac{Z_1}{Z_2} \tag{A.4*}$$

---

*Source: (Nassirharand and Karimi 2004c) © Manchester University Press

A. Nassirharand, *Computer-aided Nonlinear Control System Design: Using Describing Function Models*, DOI 10.1007/978-1-4471-2149-7,
© Springer-Verlag London Limited 2012

Fig. A.1 The structure of the considered feedback system. Source: (Nassirharand and Karimi 2004c) © Manchester University Press

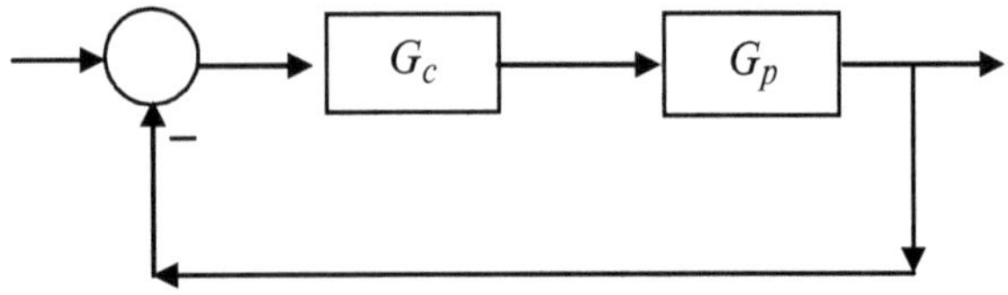

Then, the exact model matching objective function is formed:

$$J' = \int_{\omega_1}^{\omega_2} \left| \frac{Z_1}{Z_2} - \frac{Y_1}{Y_2} \right|^2 d\omega \tag{A.5*}$$

Minimizing the above function, with respect to the compensator parameters, results in a set of nonlinear algebraic equations that must be solved numerically. However, the goal is to develop a closed-form solution. Notice that under optimality conditions $\frac{Z_1}{Z_2} = \frac{Y_1}{Y_2} \Rightarrow Z_1Y_2 - Z_2Y_1 = 0$. Therefore, assuming that a solution exists, the following alternate objective function may be selected:

$$J = \int_{\omega_1}^{\omega_2} |Z_1Y_2 - Z_2Y_1|^2 d\omega \tag{A.6*}$$

By substituting Eqs. A.1–A.4 into Eq. A.6, the following is obtained:

$$J = \int_{\omega_1}^{\omega_2} \begin{bmatrix} \left(a_0A + a_1B\omega - a_2A\omega^2 + b_1D\omega + b_2E\omega^2 - E\right)^2 + \\ \left(-a_0B + a_1A\omega + a_2B\omega^2 - b_1E\omega + b_2D\omega^2 - D\right)^2 \end{bmatrix} d\omega \tag{A.7*}$$

where

$$A = h_3g_1 - h_4g_2 - h_1g_1 + h_2g_2 \tag{A.8*}$$

$$B = -h_3g_2 - h_4g_1 + h_1g_2 + h_2g_1 \tag{A.9*}$$

$$D = h_1g_4 + h_2g_3 \tag{A.10*}$$

$$E = h_1g_3 - h_2g_4 \tag{A.11*}$$

Now by taking the derivative of the objective function with respect to the compensator parameters, and by setting the result equal to zero, a set of linear

*Source: (Nassirharand and Karimi 2004c) © Manchester University Press

simultaneous algebraic equations is obtained as follows:

$$\begin{bmatrix} T_0 & 0 & -T_2 & R_1 & S_2 \\ 0 & T_2 & 0 & -S_2 & R_3 \\ -T_2 & 0 & T_4 & -R_3 & -S_4 \\ R_1 & -S_2 & -R_3 & Q_2 & 0 \\ S_2 & R_3 & -S_4 & 0 & Q_4 \end{bmatrix} \cdot \begin{bmatrix} a_0 \\ a_1 \\ a_2 \\ b_1 \\ b_2 \end{bmatrix} = \begin{bmatrix} S_0 \\ R_1 \\ -S_2 \\ 0 \\ Q_2 \end{bmatrix} \tag{A.12*}$$

where

$$T_k = \int_{\omega_1}^{\omega_2} \left(A^2 + B^2\right) \omega^k d\omega \tag{A.13*}$$

$$R_k = \int_{\omega_1}^{\omega_2} (AD + BE) \, \omega^k d\omega \tag{A.14*}$$

$$S_k = \int_{\omega_1}^{\omega_2} (AE - BD) \, \omega^k d\omega \tag{A.15*}$$

$$Q_k = \int_{\omega_1}^{\omega_2} \left(D^2 + E^2\right) \omega^k d\omega \tag{A.16*}$$

Therefore, by supplying the linear plant, the desired closed-loop transfer function, and the frequency range of interest, one could use Eq. A.12 and directly solve for the compensator parameters.

A.1.1 Software

A MATLAB® function is available that automates the above design process. The MATLAB® function is of the following form:

$$[gc] = leadlag(gp, f, w1, w2)$$

The printout of the software is given in Fig. A.2. The function inputs are (1) the linear plant transfer function, *gp*, for which a compensator is to be designed; (2) the desired closed-loop transfer function, *f*; (3) the lower bound of the frequency range of interest, *w*1; and (4) the upper bound of the frequency range of interest, *w*2. The function output is the designed lead-lag compensator, *gc*.

*Source: (Nassirharand and Karimi 2004c) © Manchester University Press

A.1.2 Demonstration Example Problem

Consider the example problem studied by Ogata (1990). The linear plant is of the following form:

```
% function leadlag designs a leadlag compensator given
% a linear plant, a desired transfer function, and a
% frequency range of interest
  function gc = leadlag(gp,f,w1,w2)
% get the numerator and denominator polynomials of the
% plant and the desired transfer function
  [gp1 gp2]=tfdata(gp,'v');[f1 f2]=tfdata(f,'v');
% set the integration step size and discretize the
% frequency range of interest
  h=0.01;w=w1:h:w2;
% given the plant and the desired transfer function, solve
% for the parameters of the
% compensator
  [x,z]=gxzkk(w,gp1,gp2,f1,f2);y=x\z;
% Form the compensator transfer function
  gc1=[y(3) y(2) y(1)];gc2=[y(5) y(4) 1];gc=tf(gc1,gc2);
% function gxzkk gets the matrix of coefficients and the
% input vector
% see Equation (A.12)
  function [x,z]=gxzkk(w,gp1,gp2,f1,f2)
% compute a, b, d, and e using Equations (A.8)-(A.11)
  [a,b,d,e]=abdekk(w,gp1,gp2,f1,f2);
% compute t, r, s, and q using Equations (A.13)-(A.16)
  t=kktk(a,b,w);r=kkrk(a,b,d,e,w);
  s=kksk(a,b,d,e,w);q=kkqk(d,e,w);
% form the matrix of coefficients using Equation (A.12)
  x=[t(1) 0      -t(3)   r(1)     s(3);
      0    t(3)    0     -s(3)    r(3);
    -t(3)  0       t(5)  -r(3)     -s(5);
     r(1)  -s(3)  -r(3)    q(2)      0;
     s(3)   r(3)   -s(5)  0       q(4)];
% form the input vector using Equation (A.12)
  z=[s(1)  r(1)  -s(3)  0  q(2)]';
```

Fig. A.2 Listing of the *leadlag* MATLAB® function. Source: (Nassirharand and Karimi 2004c) © Manchester University Press

```
% function abdekk computes a, b, d, e, given by Equations
% (A.8)-(A.11)
  function [a,b,d,e]=abdekk(w,gp1,gp2,f1,f2)
% form the complex frequency s
  s=sqrt(-1)*w;
% separate the real and imaginary parts of the numerator
% and  denominator polynomials of the known transfer
% functions see Equations (A.1) and (A.3)
  fsn1=polyval(gp1,s);g1=real(fsn1);g2=imag(fsn1);
  fsp2=polyval(gp2,s);g3=real(fsp2);g4=imag(fsp2);
  fshd1=polyval(f1,s);h1=real(fshd1);h2=imag(fshd1);
  fshd2=polyval(f2,s);h3=real(fshd2);h4=imag(fshd2);
% use Equations (A.8)-(A.11) and compute a, b, d, and e
  a=h3.*g1-h4.*g2-g1.*h1+h2.*g2;b=-h3.*g2-
h4.*g1+h1.*g2+h2.*g1;
  d=h1.*g4+h2.*g3;e=h1.*g3-h2.*g4;
% function t computes array t given by Equation (A.13)
  function t = kktk(a,b,w)
% use Equation (A.13) and compute array t
  for k=1:2:5;k1=k-
1;y=(w.^k1).*(a.^2+b.^2);t(k)=trapz(w,y);end
% function r computes array r given by Equation (A.14)
  function r = kkrk(a,b,d,e,w)
% use Equation (A.14) and compute array r
  for k=1:2:3;y=(w.^k).*(a.*d+b.*e);r(k)=trapz(w,y);end
% function s computes array s given by Equation (A.15)
  function s = kksk(a,b,d,e,w)
% compute array s using Equation (A.15)
  for k=1:2:5;kk=k-1; y=(w.^kk).*(a.*e-b.*d);
s(k)=trapz(w,y);end
% function q computes array q given by Equation (A.16)
  function q = kkqk(d,e,w)
% compute array q using Equation (A.16)
for k=2:2:4;y=(w.^k).*(d.^2+e.^2);q(k)=trapz(w,y);end
```

Fig. A.2 (continued)

$$G_p = \frac{4}{s\,(s+0.5)} \tag{A.17}$$

It is desired to design a lead-lag controller that results in a closed-loop system that would have the damping ratio of the dominant closed-loop poles equal 0.5 with an undamped natural frequency of 5 rad/s; the static velocity error constant is to be 80 s^{-1}. The corresponding desired closed-loop transfer function is obtained as follows Chen and Shieh (1970). The desired transfer function is of the following form:

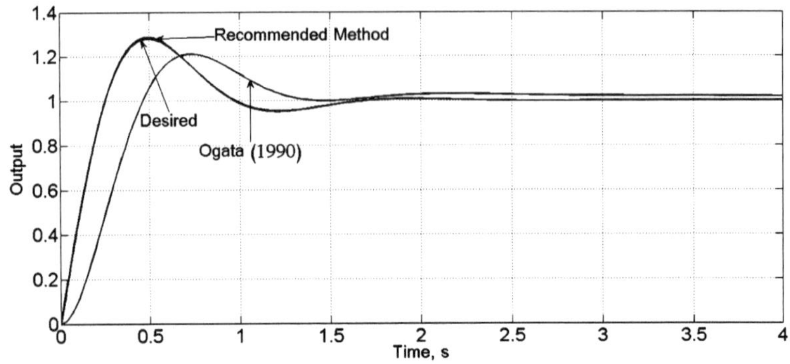

Fig. A.3 Design comparison for linear lead-lag design example. Source: (Nassirharand and Karimi 2004c) © Manchester University Press

$$F = \frac{xs + \omega_n^2}{s^2 + 2\varsigma\omega_n s + \omega_n^2} \tag{A.18}$$

where from the problem statement, $\omega_n = 5$and $\varsigma = 0.5$. Based on the definition of velocity error constant, k_v, the following relation holds:

$$2\varsigma\omega_n - \left(\frac{\omega_n^2}{k_v}\right) - x = 0 \quad \Rightarrow x = 4.6875 \tag{A.19}$$

Hence,

$$F = \frac{4.6875s + 25}{s^2 + 5s + 25} \tag{A.20}$$

The developed MATLAB® function is executed with $w1 = 0$ and $w2 = 5$:

$$gc = leadlag\,(Gp, F, 0, 5)$$

The designed compensator is of the following form:

$$G_c = \frac{3.75s^2 + 21.875s + 10}{\varepsilon s^2 + 3.2s + 1}.Let\varepsilon = 0.01. \tag{A.21}$$

The design given by Ogata 1990 using a classical technique is of the following form:

$$C_{ogata}(s) = \frac{10\,(2s + 1)\,(5s + 1)}{(0.1992s + 1)\,(80.19s + 1)} \tag{A.22}$$

The performance of the two compensators may be compared by studying Fig. A.3.

A.2 Linear PID Controller Design

With a similar formulation for the linear lead-lag controller design that is presented in the previous section, the gains of a PID controller may be determined to be as follows (Nassirharand et al. 2003):

$$K_P = \frac{Q}{T_2} \tag{A.23*}$$

$$\begin{bmatrix} T_0 & -T_2 \\ -T_2 & T_4 \end{bmatrix} \begin{bmatrix} K_I \\ K_D \end{bmatrix} = \begin{bmatrix} -R \\ S \end{bmatrix} \tag{A.24*}$$

where

$$T_l = \int_{\omega_1}^{\omega_2} \omega^l \left(A^2 + B^2\right) d\omega, \quad l = 0, 2 \tag{A.25*}$$

$$R = \int_{\omega_1}^{\omega_2} (AC + BD)\, d\omega \tag{A.26*}$$

$$S = \int_{\omega_1}^{\omega_2} (AC + BD)\, \omega^2 d\omega \tag{A.27*}$$

$$Q = \int_{\omega_1}^{\omega_2} (BC - AD)\, \omega\, d\omega \tag{A.28*}$$

$$A = g_1 h_3 - g_2 h_4 - g_1 h_1 + g_2 h_2 \tag{A.29*}$$

$$B = g_2 h_3 + g_1 h_4 - g_2 h_1 - g_1 h_2 \tag{A.30*}$$

$$C = \omega\,(g_4 h_1 + g_3 h_2) \tag{A.31*}$$

$$D = \omega\,(g_4 h_2 - g_3 h_1) \tag{A.32*}$$

$$G_c = K_P + K_I/s + K_D s \tag{A.33*}$$

and the rest of the variables are defined as the previous section (see Eqs. A.1 and A.3).

*Source: (Nassirharand et al. 2003) © Elsevier

Therefore, by supplying the linear plant, the desired closed-loop transfer function, and the frequency range of interest, one could use Eqs. A.23 and A.24 and directly solve for the controller parameters.

A.2.1 Software

The above described method is automated via the development of a new MATLAB® function. The developed MATLAB® function is of the following form:

$$[kp, ki, kd] = getpid\,(w1, w2, ss, g, hd)$$

Where in the above command, $[kp, ki, kd]$ are the controller gains; the inputs are (1) $w1, w2$ the frequency range of interest, (2) ss the step size for integration of Eqs. A.25–A.28, (3) g the transfer function of the process, and (4) h_d the transfer function of the reference linear model. The software listing is shown in Fig. A.4.

A.2.2 Demonstration Example Problem

Consider the following first-order plus delay model studied by Wang et al. (1995):

$$G_p(s) = \frac{e^{-5s}}{10s + 1} \tag{A.34*}$$

The desired transfer function is of the following form:

$$F(s) = \frac{e^{-5s}}{20s + 1} \tag{A.35*}$$

Notice that the developed procedure must have all transfer functions as rational functions in s; for this reason, we approximate the e^{-5s} term as below (Kuo 1982):

$$e^{-Ts} \approx \frac{1}{[1 + Ts/n]^n} \tag{A.36*}$$

where $T = 5$, and we arbitrarily select the order of approximation as $n = 4$. By substituting Eq. A.36 into Eqs. A.34 and A.35, the developed MATLAB® function is executed with $w1 = 0$, $w2 = 0.025$, and $h = 0.001$. Using Wang et al. (1995), the results are compared with four other techniques in Fig. A.5, and the corresponding values of PID controller gains are tabulated in Table A.1. It is apparent that, in this case, the developed approach as well as method given by Wang et al. (1995) produces better results than other considered techniques.

---

*Source: (Nassirharand et al. 2003) © Elsevier

A.3 Proportional Plus Rate Feedback Controller Design

Consider the feedback system depicted in Fig. A.6. In this Appendix, the formulation for design of proportional plus rate feedback controllers is presented (Nassirharand 2011). The linear plant is expressed as follows:

$$G_p(j\omega) = A(\omega) + jB(\omega) \tag{A.37*}$$

```
% function getpid designs a PID controller given
% a linear plant, a desired transfer function, and a
% frequency range of interest
%
  function [kp,ki,kd]=getpid(w1,w2,h,g,hd)
% discretize frequency and extract the numerator
% and denominator polynomials
  w=w1:h:w2;[hd1,hd2]=tfdata(hd,'v');
  [g1,g2]=tfdata(g,'v');
% solve for PID gains
  [x,z,kp]=gxzpid(w,g1,g2,hd1,hd2);
% x*y=Z
  y=x\z(:); see Equation (A.24)
  ki=y(1);kd=y(2);
%
% function gxzpid determines the x and z noted above
% and the gain kp from  Equation (A.23)
  function [x,z,kp] = gxzpid (w,g1,g2,hd1,hd2)
% compute a, b, c, d given by equations (A.29)-(A.32)
  [a,b,c,d]=abcdpid(w,g1,g2,hd1,hd2);
  % compute t (see Equation A.25)
  t=tlpid(a,b,w);
% compute r (see Equation A.26)
  r=rlpid(a,b,c,d,w);
% compute s (see equation A.27)
  s=slpid(a,b,c,d,w);
%Compute q (see Equation A.28)
  q=qlpid(a,b,c,d,w);
```

Fig. A.4 Listing for PID design function. Source: (Nassirharand et al. 2003) © Elsevier

*Source: (Nassirharand 2011) © Elsevier

```
% form x and z (see Equation A.24)
  x(1,1)=t(1);x(1,2)=-t(3);x(2,1)=-t(3);x(2,2)=t(4);
  z(1)=-r;z(2)=s;
%Compute kp (see Equation A.23)
  kp=q/t(3);
%
% function abcdpid computes a, b, c, and d
% see Equations (A.29) - (A.32)
  function [a,b,c,d]=abcdpid(w,g1,g2,hd1,hd2)
% compute complex frequency
  s=sqrt(-1)*w;
% see Equation (A.1)→ sa=g1, sb=g2, sc=g3, sd=g4
  fsg1=polyval(g1,s);sa=real(fsg1);sb=imag(fsg1);
  fsg2=polyval(g2,s);sc=real(fsg2);sd=imag(fsg2);
% see Equation (A.3) → e=h1, f=h2, sg=h3, sh=h4
  fshd1=polyval(hd1,s);e=real(fshd1);f=imag(fshd1);
  fshd2=polyval(hd2,s);sg=real(fshd2);sh=imag(fshd2);
% compute a from Equation (A.29)
  a=sa.*sg-sb.*sh-sa.*e+sb.*f;
% compute b from Equation (A.30)
  b=sb.*sg+sa.*sh-sb.*e-sa.*f;
% Compute c from Equation (A.31)
  c=w.*(sd.*e+sc.*f);
% compute d from Equation (A.32)
  d=w.*(sd.*f-sc.*e);
% function tlpid computes t from Equation (A.25)
  function t = tlpid(a,b,w)
  for l=1:2:5
     l1=l-1;y=(w.^l1).*(a.^2+b.^2);
     t(l)=trapz(w,y);
  end
% function rlpid computes r from Equation (A.26)
  function r = rlpid(a,b,c,d,w)
  y=a.*c+b.*d;r=trapz(w,y);
% function slpid computes s from Equation (A.27)
  function s = slpid(a,b,c,d,w)
  y=(w.^2).*(b.*d+a.*c);s=trapz(w,y);
% function qlpid computes q from Equation (A.28)
  function q = qlpid(a,b,c,d,w)
  y=w.*(b.*c-a.*d);q=trapz(w,y);
```

Fig. A.4 (continued)

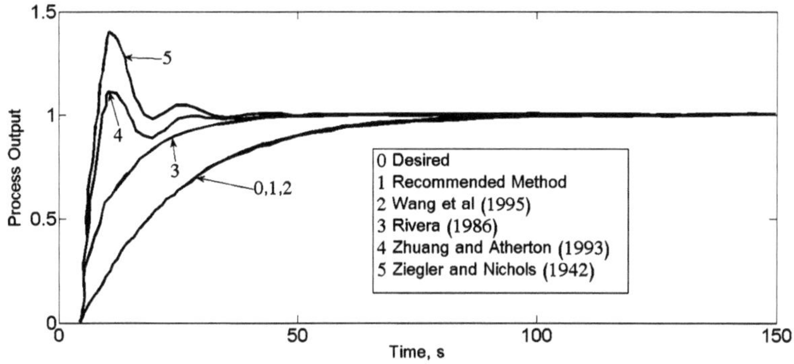

Fig. A.5 Design comparison for PID controller design example. Source: (Nassirharand et al. 2003) © Elsevier

Table A.1 Comparison of PID gains with different design methods – (1) Wang et al. (1995), (2) Rivera et al. (1986), (3) Ziegler and Nichols (1942), and (4) Zhuang and Atherton (1993)

	Present	(1)	(2)	(3)	(4)
K_p	0.42	0.42	1.00	2.28	1.99
K_I	0.04	0.04	0.08	0.26	0.13
K_D	0.30	0.17	2.00	4.89	3.84

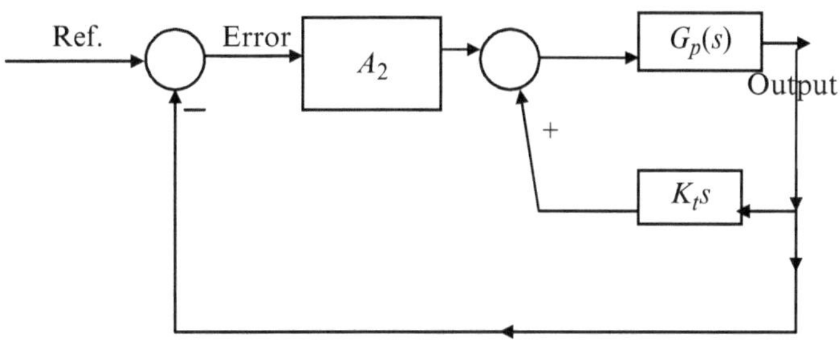

Fig. A.6 The structure of the proportional and rate feedback controller. Source: (Nassirharand 2011) © Elsevier

The desired closed-loop system behavior, or the desired reference linear model, is denoted F:

$$F\left(j\omega\right)=C\left(\omega\right)+jD\left(\omega\right) \tag{A.38*}$$

The designed closed-loop system is denoted G_{cl}:

$$G_{cl}=\frac{A\left(\omega\right)A_2+jB\left(\omega\right)A_2}{\left(1+K_t\omega B\left(\omega\right)+A\left(\omega\right)A_2\right)+j\left(-K_tA\left(\omega\right)\omega+A_2B\left(\omega\right)\right)}=\frac{X_1\left(j\omega\right)}{X_2\left(j\omega\right)} \tag{A.39*}$$

*Source: (Nassirharand 2011) © Elsevier

Then, the exact model matching objective function is formed:

$$J' = \int_{\omega_1}^{\omega_2} \left| \frac{X_1(j\omega)}{X_2(j\omega)} - C(\omega) - jD(\omega) \right|^2 d\omega \tag{A.40*}$$

Minimizing the above function, with respect to the compensator parameters, results in a set of nonlinear algebraic equations that must be solved numerically. However, the goal is to develop a closed-form solution. Notice that under optimality conditions.

$$\frac{X_1(j\omega)}{X_2(j\omega)} = C(\omega) + jD(\omega) \Rightarrow X_1(j\omega) - X_2(j\omega)C - jX_2(j\omega)D(\omega) = 0$$

Therefore, assuming that a solution exists, the following alternate objective function may be selected:

$$J = \int_{\omega_1}^{\omega_2} |X_1(j\omega) - X_2(j\omega)C - jX_2(j\omega)D(\omega)|^2 d\omega \tag{A.41*}$$

By comparing Eqs. A.35 and A.37, the following is obtained:

$$J = \int_{\omega_1}^{\omega_2} \left[\begin{pmatrix} A(\omega)A_2 - C(\omega) - C(\omega)K_t\omega B(\omega) - C(\omega)A(\omega)A_2 \\ -D(\omega)K_tA(\omega)\omega - D(\omega)B(\omega)A_2 + B(\omega)A_2 \\ +K_tA(\omega)\omega C(\omega) - A_2B(\omega)C(\omega) - D(\omega) \\ -D(\omega)K_t\omega B(\omega) - D(\omega)A(\omega)A_2 \end{pmatrix}^2 \right] d\omega \tag{A.42*}$$

Now by taking the derivative of the objective function with respect to the controller parameters, and by setting the result equal to zero, a set of linear simultaneous algebraic equations is obtained as follows:

$$\begin{bmatrix} \alpha_1 & \beta_1 \\ \alpha_2 & \beta_2 \end{bmatrix} \begin{bmatrix} K_t \\ A_2 \end{bmatrix} = \begin{bmatrix} -\gamma_1 \\ -\gamma_2 \end{bmatrix} \tag{A.43*}$$

where

$$\alpha_1 = \int_{\omega_1}^{\omega_2} \left[-D(\omega)\omega \left(A^2(\omega) + B^2(\omega) \right) \right] d\omega \tag{A.44*}$$

---

*Source: (Nassirharand 2011) © Elsevier

$$\beta_1 = \int_{\omega_1}^{\omega_2} \left\{ \left(A^2(\omega) + B^2(\omega)\right) \left[(1 - C(\omega))^2 + D^2(\omega) \right] \right\} d\omega \qquad \text{(A.45*)}$$

$$\alpha_2 = \int_{\omega_1}^{\omega_2} \left[\omega^2 \left(A^2(\omega) + B^2(\omega)\right) \left(C^2(\omega) + D^2(\omega)\right) \right] d\omega \qquad \text{(A.46*)}$$

$$\beta_2 = \alpha_1 \qquad \text{(A.47*)}$$

$$\gamma_1 = \int_{\omega_1}^{\omega_2} \left[A(\omega) \left(C^2(\omega) + D^2(\omega)\right) - A(\omega) C(\omega) - B(\omega) D(\omega) \right] d\omega \qquad \text{(A.48*)}$$

$$\gamma_2 = \int_{\omega_1}^{\omega_2} \left[\omega B(\omega) \left(C^2(\omega) + D^2(\omega)\right) \right] d\omega \qquad \text{(A.49*)}$$

Therefore, by supplying the linear plant, the desired closed-loop transfer function, and the frequency range of interest, one could use Eq. A.43 and directly solve for the controller parameters.

A.3.1 Software

The design process is automated via development of a new MATLAB® function. The MATLAB® function is of the following form:

$$[Kt, A2] = get\_L\_rate(w1, w2, h, g, f)$$

The function inputs are (1) the lower bound of the frequency range of interest, $w1$; (2) the upper bound of the frequency range of interest, $w2$; (3) the integration step size; (4) the linear plant transfer function, g, for which a compensator is to be designed; and (5) the desired closed-loop transfer function, f. The function outputs are the tachometer gain and the amplifier gain. The corresponding software listing is given in Fig. A.7 (Nassirharand 2011).

The design process is automated via development of a new MATLAB® function. The MATLAB® function is of the following form:

$$[Kt, A2] = get\_L\_rate(w1, w2, h, g, f)$$

*Source: (Nassirharand 2011) © Elsevier

The function inputs are (1) the lower bound of the frequency range of interest, $w1$; (2) the upper bound of the frequency range of interest, $w2$; (3) the integration step size; (4) the linear plant transfer function, g, for which a compensator is to be designed; and (5) the desired closed-loop transfer function, f. The function outputs are the tachometer gain and the amplifier gain. The corresponding software listing follows.

```
function [Kt,A2]=get_L_rate(w1,w2,h,g,f)
%this function determines the required gains of the
%  controller
%discretize frequency range; get numerator and
%denominator polynomials of
%desired and plant transfer functions
w=w1:h:w2;[f1,f2]=tfdata(hd,'v');[g1,g2]=tfdata(g,'v');
%obtain the coefficient matrix x and the input vector z
% (see Equation (A.43)
[x,z]=gxzLrate(w,g1,g2,f1,f2);
%solve matrix Equation (A.43)
y=x\z(:);
%get the gains
Kt=y(1);A2=y(2);
%%%%%%%%
function [x,z] = gxzLrate (w,g1,g2,f1,f2)
%this function determines the coefficient matrix x and
% the input vector z
%A, B, C, D are defined via Equations (A.37) and (A.38)
[A,B,C,D]=abcdLrate(w,g1,g2,f1,f2);
%determine alpha1; see Equation (A.44)
aux1=-D.*w;A2B2=A.^2+B.^2;y=aux1.*A2B2;
alpha1=trapz(w,y)
%determine beta 1; see Equation (A.45)
one=ones(size(C));y=((one-C).^2+D.^2).*A2B2;
beta1=trapz(w,y);
%determine gamma1; see Equation (A.48)
C2D2=C.^2+D.^2;y=A.*C2D2-A.*C-B.*D;
gamma1=trapz(w,y)
%determine alpha2; see Equation (A.46)
aux1=w.^2;aux2=A2B2.*C2D2;y=aux1.*aux2;
alpha2=trapz(w,y)
%determine beta2; see Equation (A.47)
beta2=alpha1;
```

Fig. A.7 The software listing for proportional and rate feedback design. Source: (Nassirharand 2011) © Elsevier

```
  %see gamma2; see Equation (A.49)
  aux1=w.*B;y=aux1.*C2D2;
  gamma2=trapz(w,y)
  %form the coefficient matrix x
  x(1,1)=alpha1;x(1,2)=beta1;x(2,1)=alpha2;x(2,2)=beta2
;
  %form the input vector z
  z(1)=-gamma1;z(2)=-gamma2;
  %%%%%%
  function [A,B,C,D]=abcdLrate(w,g1,g2,f1,f2)
  %this function determines A, B, C, and D based on
  % Equations (A.37) and (A.38)
  s=sqrt(-1)*w;
  fsg1=polyval(g1,s);fsg2=polyval(g2,s);
  aux=fsg1./fsg2;
  A=real(aux);B=imag(aux);
  fsf1=polyval(f1,s);fsf2=polyval(f2,s);
  aux=fsf1./fsf2;
  C=real(aux);D=imag(aux);
%end
```

Fig. A.7 (continued)

A.3.2 Demonstration Example Problem

Consider the example problem studied by D'Azzo and Houpis (1966). The linear plant is of the following form:

$$G_p = \frac{0.88}{s(s+1)(0.2s+1)} \tag{A.50*}$$

It is desired to design an amplifier and rate feedback controller that results in a closed-loop system that would have $M_m = 1.25$ and $\omega_m = 3$. These performance measures correspond to a desired second-order transfer function with $\varsigma = 0.4472$ and $\omega_n = 3.873$; i.e.:

$$F = \frac{15}{s^2 + 3.464s + 15} \tag{A.51*}$$

In arriving at the above desired transfer function, the following formulas were utilized:

$$M_m = \frac{1}{2\varsigma\sqrt{1-\varsigma^2}} \tag{A.52*}$$

*Source: (Nassirharand 2011) © Elsevier

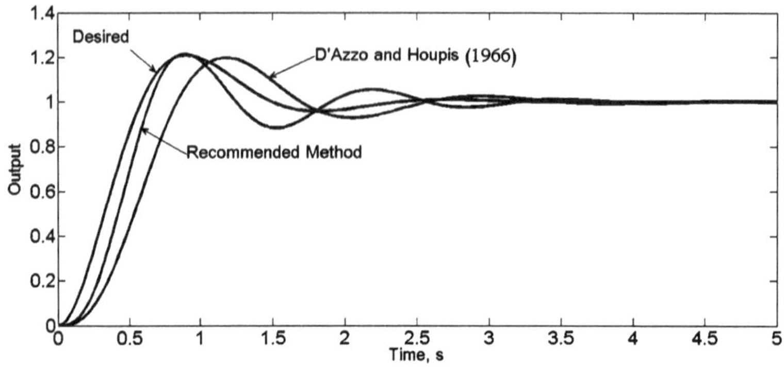

Fig. A.8 Design comparison for the proportional and rate feedback controller example. Source: (Nassirharand 2011) © Elsevier

$$\omega_m = \omega_n \sqrt{1 - 2\varsigma^2} \tag{A.53*}$$

The developed MATLAB function is executed with $w1 = 2.5$ and $w2 = 10$ as follows:

$$[Kt, A2] = get\_L\_rate\,(2.5, 10, 0.01, gp, f)$$

The designed gains are $K_t = -6.3557$ and $A_2 = 20.4545$.

The designed gains using a classical technique are $K_t = -4$ and $A_2 = 11.4$ (D'Azzo and Houpis 1966).

The performance of the two compensators may be compared by studying Fig. A.8. It should be noted that by selecting F the same as the closed-loop transfer function with gains of D'Azzo and Houpis (1966), then the present approach also gives the same exact gains.

A.4 Proportional-Integral Plus Rate Feedback Controller Design

The structure of the considered feedback system is depicted in Fig. A.9.

The problem statement is to design a PI plus rate feedback compensator for a linear plant (as shown in Fig. A.9) that would satisfy a set of user-defined performance measures. The formulation for PI plus rate feedback follows. The linear plant is expressed as follows (Nassirharand 2011):

*Source: (Nassirharand 2011) © Elsevier

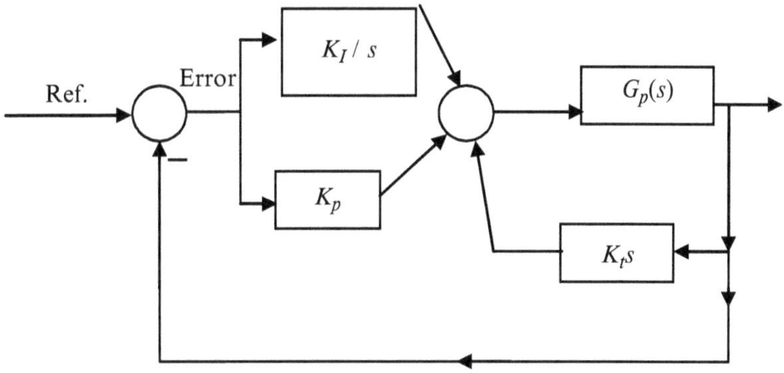

Fig. A.9 Structure of the proportional-integral plus rate feedback controller. Source: (Nassirharand 2011) © Elsevier

$$G_p(j\omega) = A(\omega) + jB(\omega) \tag{A.54*}$$

The desired closed-loop system behavior, or the desired reference linear model, is denoted F:

$$F(j\omega) = C(\omega) + jD(\omega) \tag{A.55*}$$

The designed closed-loop system is denoted G_{cl}:

$$\begin{aligned} G_{cl} &= \frac{(AK_I - K_P B\omega) + j(BK_I + AK_P\omega)}{(AK_t\omega^2 + AK_I - K_P B\omega) + j(\omega + BK_t\omega^2 + BK_I + AK_P\omega)} \\ &= \frac{X_1(j\omega)}{X_2(j\omega)} \end{aligned} \tag{A.56*}$$

Then, the exact model matching objective function is formed:

$$J' = \int_{\omega_1}^{\omega_2} \left| \frac{X_1(j\omega)}{X_2(j\omega)} - C(\omega) - jD(\omega) \right|^2 d\omega \tag{A.57*}$$

Minimizing the above function, with respect to the compensator parameters, results in a set of nonlinear algebraic equations that must be solved numerically. However, the goal is to develop a closed-form solution. Notice that under optimality conditions

*Source: (Nassirharand 2011) © Elsevier

$$\frac{X_1(j\omega)}{X_2(j\omega)} = C(\omega) + jD(\omega) \Rightarrow X_1(j\omega) - X_2(j\omega)C - jX_2(j\omega)D(\omega) = 0.$$

Therefore, assuming that a solution exists, the following alternate objective function may be selected:

$$J = \int_{\omega_1}^{\omega_2} |X_1(j\omega) - X_2(j\omega)C - jX_2(j\omega)D(\omega)|^2 d\omega \qquad \text{(A.58*)}$$

By comparing Eqs.A.50 and A.54, the following is obtained:

$$J = \int_{\omega_1}^{\omega_2} \left[\begin{array}{l} \left(\begin{array}{l} AK_I - K_P B\omega - ACK_t\omega^2 - ACK_I + K_P CB\omega \\ +\omega D + BDK_t\omega^2 + BDK_I + ADK_P\omega \end{array} \right)^2 \\ + \left(\begin{array}{l} BK_I + AK_P\omega - DAK_t\omega^2 - ADK_I + K_P DB\omega \\ -\omega C - BCK_t\omega^2 - BCK_I - ACK_P\omega \end{array} \right)^2 \end{array} \right] d\omega \qquad \text{(A.59*)}$$

Now by taking the derivative of the objective function with respect to the compensator parameters, and by setting the result equal to zero, a set of linear simultaneous algebraic equations is obtained as follows:

$$XY = Z \qquad \text{(A.60*)}$$

where

$$X = \begin{bmatrix} \alpha_1 & \beta_1 & \gamma_1 \\ \alpha_2 & \beta_2 & \gamma_2 \\ \alpha_3 & \beta_3 & \gamma_3 \end{bmatrix} \qquad \text{(A.61*)}$$

$$Y = \begin{bmatrix} K_I \\ K_P \\ K_t \end{bmatrix} \qquad \text{(A.62*)}$$

$$Z = \begin{bmatrix} -\eta_1 \\ -\eta_2 \\ -\eta_3 \end{bmatrix} \qquad \text{(A.63*)}$$

$$\alpha_1 = \int_{\omega_1}^{\omega_2} \left\{ \left(A^2 + B^2\right) \left[(C-1)^2 + D^2 \right] \right\} d\omega \qquad \text{(A.64*)}$$

$$\beta_1 = 0 \qquad \text{(A.65*)}$$

*Source: (Nassirharand 2011) © Elsevier

$$\gamma_1 = \int_{\omega_1}^{\omega_2} \left\{\omega^2 \left(A^2 + B^2\right)\left(C^2 + D^2 - C\right)\right\} d\omega \qquad \text{(A.66*)}$$

$$\eta_1 = \int_{\omega_1}^{\omega_2} \left\{B\omega \left(C^2 + D^2\right) + \omega \left(AD - BC\right)\right\} d\omega \qquad \text{(A.67*)}$$

$$\alpha_2 = 0 \qquad \text{(A.68*)}$$

$$\beta_2 = \int_{\omega_1}^{\omega_2} \left\{\omega^2 \left(A^2 + B^2\right)\left[(C-1)^2 + D^2\right]\right\} d\omega \qquad \text{(A.69*)}$$

$$\gamma_2 = \int_{\omega_1}^{\omega_2} \left\{-\omega^3 D \left(A^2 + B^2\right)\right\} d\omega \qquad \text{(A.70*)}$$

$$\eta_2 = \int_{\omega_1}^{\omega_2} \left\{\omega^2 \left[A\left(C^2 + D^2\right) - (AC + BD)\right]\right\} d\omega \qquad \text{(A.71*)}$$

$$\alpha_3 = \int_{\omega_1}^{\omega_2} \left\{\omega^2 \left[C(C-1) + D^2\right]\left(A^2 + B^2\right)\right\} d\omega \qquad \text{(A.72*)}$$

$$\beta_3 = \int_{\omega_1}^{\omega_2} \left\{-\omega^3 D \left(A^2 + B^2\right)\right\} d\omega \qquad \text{(A.73*)}$$

$$\gamma_3 = \int_{\omega_1}^{\omega_2} \left\{\omega^4 \left(A^2 + B^2\right)\left(C^2 + D^2\right)\right\} d\omega \qquad \text{(A.74*)}$$

$$\eta_3 = \int_{\omega_1}^{\omega_2} \left\{\omega^3 B \left(C^2 + D^2\right)\right\} d\omega \qquad \text{(A.75*)}$$

Therefore, by supplying the linear plant, the desired closed-loop transfer function, and the frequency range of interest, one could use Eq. A.60 and directly solve for the compensator gains.

---

*Source: (Nassirharand 2011) © Elsevier

A.4.1 Software

The design process is automated via development of a new MATLAB® function. The MATLAB® function is of the following form:

$$[KI, KP, Kt] = get\_PIL\_rate(w1, w2, h, g, f)$$

The function inputs are (1) the lower bound of the frequency range of interest, $w1$; (2) the upper bound of the frequency range of interest, $w2$; (3) the integration step size, h; (4) the linear plant transfer function, g, for which a compensator is to be designed; and (5) the desired closed-loop transfer function, f. The function outputs are the integral, proportional, and tachometer gains. The corresponding software listing is given in Fig. A.10 (Nassirharand 2011).

A.4.2 Demonstration Example Problem

Consider the example problem studied by D'Azzo and Houpis (1966). The linear plant is of the following form:

$$G_p = \frac{0.88}{s\,(s+1)\,(0.2s+1)} \tag{A.76*}$$

It is desired to design a PI plus rate feedback compensator that results in a closed-loop system that would have $M_m = 1.25$ and $\omega_m = 3$. These performance measures correspond to a desired second-order transfer function with $\varsigma = 0.4472$ and $\omega_n = 3.873$; i.e.:

$$F = \frac{15}{s^2 + 3.464s + 15} \tag{A.77*}$$

The developed MATLAB® function is executed with $w1 = 2.5$ and $w2 = 10$ as follows:

$$[KI, KP, Kt] = get\_PIL\_rate\,(0.001, 6, 0.01, Gp, F)$$

The designed gains are $K_I = 0$, $K_P = 20.45$, and $K_t = -5.74$.

The gains designed by D'Azzo and Houpis (1966) using a classical technique for only proportional and rate feedback are $K_t = -4$ and $K_p = 11.4$.

The performance of the two compensators may be compared by studying Fig. A.11.

---

*Source: (Nassirharand 2011) © Elsevier

```
function [KI,KP,Kt]=get_PIL_rate(w1,w2,h,g,hd)
%The purpose of this function is to compute the gains
% KI, KP, and Kt
%The inputs are defined in the paper
w=w1:h:w2;[hd1,hd2]=tfdata(hd,'v');[g1,g2]=tfdata(g,'v');
%get x and z in Equations (A.61) and (A.63)
[x,z]=gxzPILrate(w,g1,g2,hd1,hd2);
%From Equation (A.60) solve for the gains; y is given by
% equation (A.62)
y=x\z(:);
KI=y(1);KP=y(2);Kt=y(3);
%
function [x,z] = gxzPILrate (w,g1,g2,hd1,hd2)
%The purpose of this function is to form matrix x
% and vector z
%
%Evaluate A,B,C, and D given by Equations (A.54) and (A.55)
[A,B,C,D]=abcdPILrate(w,g1,g2,hd1,hd2);
%Evaluate alpha1 given by Equation (A.64)
A2B2=A.^2+B.^2;one=ones(size(C));aux1=(C-
one).^2+D.^2;y=A2B2.*aux1;alpha1=trapz(w,y);
%Set beta1 to zero according to Equation (A.65)
beta1=0;
%Evaluate gamma1 given by Equation (A.66)
clear y;
C2D2=C.^2+D.^2;aux=C2D2-
C;y=(w.^2).*(A2B2.*aux);gamma1=trapz(w,y);
%Evaluate eta1 given by Equation (A.67)
clear y;
aux1=B.*C2D2;y=(w).*(aux1+A.*D-B.*C);eta1=trapz(w,y);
%Set alpha2 to zero according to Equation (A.68)
alpha2=0;
```

Fig. A.10 Software listing for proportional-integral plus rate feedback design. Source: (Nassirharand 2011) © Elsevier

```
%Evaluate beta2 given by Equation (A.69)
clear y;
aux1=(C-one).^2+D.^2;aux3=A2B2.*aux1;
y=(w.^2).*(aux3);beta2=trapz(w,y);
%Evaluate gamma2 given by Equation (A.70)
clear y;y=((-(w.^3)).*D).*A2B2;gamma2=trapz(w,y);
%Evaluate eta2 given by Equation (A.71)
clear y;y=(w.^2).*( (A.*C2D2)-(A.*C+B.*D) );eta2=trapz(w,y);
%Evaluate alpha3 given by equation (A.72)
clear y;
aux1=C.*(C-one);
aux2=aux1+D.^2;y=(w.^2).*(A2B2.*aux2);alpha3=trapz(w,y);
%Evaluate beta3 given by Equation (A.73)
clear y;y=((-(w.^3)).*D).*A2B2;beta3=trapz(w,y);
%Evaluate gamma3 given by Equation (A.74)
clear y;y=((w.^4).*A2B2).*C2D2;gamma3=trapz(w,y);
%Evaluate eta3 given by Equation (A.75)
clear y;y=((w.^3).*B).*C2D2;eta3=trapz(w,y);
%Form matrix x according to Equation (A.61)
x(1,1)=alpha1;x(1,2)=beta1;x(1,3)=gamma1;
x(2,1)=alpha2;x(2,2)=beta2;x(2,3)=gamma2;
x(3,1)=alpha3;x(3,2)=beta3;x(3,3)=gamma3;
%Form vector z according to Equation (A.63)
z(1)=-eta1;z(2)=-eta2;z(3)=-eta3;
%
```

Fig. A.10 (continued)

Remark The MATLAB® functions, which are developed in this appendix, are integrated with the *cad_controller* MATLAB® function as described in Appendix B. With the *cad_controller* function, the specific values of $w1$ and $w2$ are determined by optimization; if f requires enhancing (see Appendix B), then that function will automatically perform enhancing.

```
function [A,B,C,D]=abcdPILrate(w,g1,g2,hd1,hd2)
%The purpose of this function is to evaluate A, B, C, and D
%given in Equations (A.54) and (A.55)
%generate the complex frequency array
s=sqrt(-1)*w;
%Generate A and B
fsg1=polyval(g1,s);fsg2=polyval(g2,s);aux=fsg1./fsg2;
A=real(aux);B=imag(aux);
%Generate C and D
fshd1=polyval(hd1,s);fshd2=polyval(hd2,s);aux=fshd1./fshd2;
C=real(aux);D=imag(aux);
```

Fig. A.10 (continued)

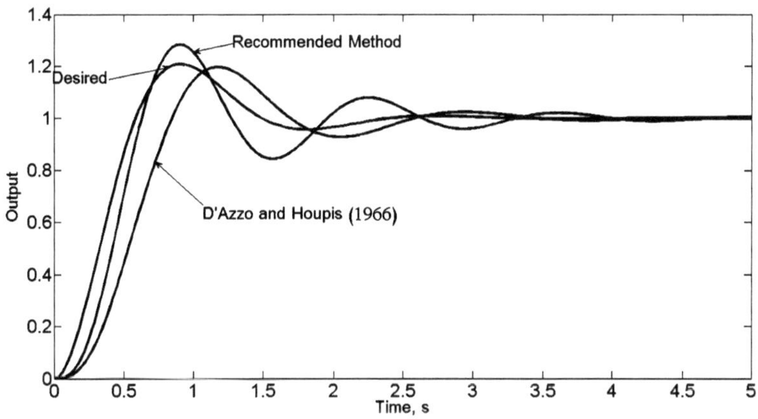

Fig. A.11 Design comparison for proportional-integral plus rate feedback controller. Source: (Nassirharand 2011) © Elsevier

Appendix B
Algebraic Linear Multivariable Controller Design

B.1 Theory

The design procedure is demonstrated for a typical two-input two-output process (Mousavi Firdeh et al. 2006). With reference to Fig. B.1, the following relations hold:

$$y_i = e_i\left(g_{ii}c_{ii} + g_{ij}c_{ji}\right) + e_j\left(g_{ii}c_{ij} + g_{ij}c_{jj}\right), \quad \{i = 1,2; j = 1,2; i \neq j\} \tag{B.1*}$$

To achieve decoupling, the following relations must hold:

$$c_{ij} = -\frac{g_{ij}c_{jj}}{g_{ii}}, \quad \{i = 1,2; j = 1,2; i \neq j\} \tag{B.2*}$$

To achieve tracking, compare Eqs. B.1 and B.2 to obtain the following:

$$\frac{y_i}{r_i} = \frac{c_{ii}k_i}{1 + c_{ii}k_i}, \quad i = 1,2 \tag{B.3*}$$

where

$$k_i = g_{ii} - \frac{g_{ij}g_{ji}}{g_{jj}}, \quad \{i = 1,2; j = 1,2; i \neq j\} \tag{B.4*}$$

In order to design c_{11} and c_{22}, a factorization approach may be utilized.

The class of all compensators that stabilize a linear plant $G(s)$ may be parameterized in terms of a stable function parameter $r(s)$ as follows (Youla et al. 1976; Vidyasagar 1985):

---

*Source: (Mousavi Firdeh et al. 2006) © Elsevier

A. Nassirharand, *Computer-aided Nonlinear Control System Design: Using Describing Function Models*, DOI 10.1007/978-1-4471-2149-7,

© Springer-Verlag London Limited 2012

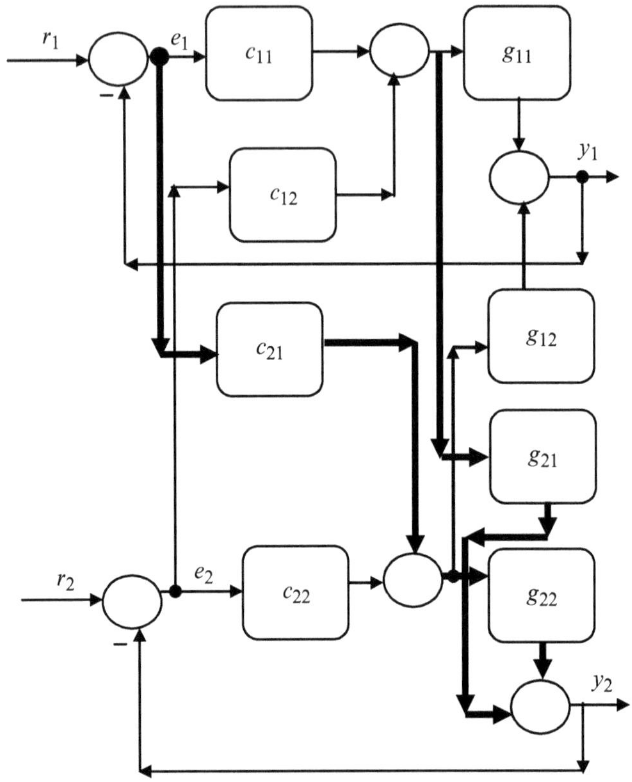

Fig. B.1 A linear two-input two-output system

$$C(s) = \frac{P(s) - r(s)D(s)}{Q(s) + r(s)N(s)} \tag{B.5*}$$

where $C(s)$ is the stabilizing compensator, $N(s)$ and $D(s)$ are stable coprime transfer function factors of the plant satisfying $G(s) = \frac{N(s)}{D(s)}$, $P(s)$ and $Q(s)$ are stable coprime transfer functions and satisfy the Bezout's identity: $P(s)N(s) + Q(s)D(s) = 1$, and $r(s)$ is an as yet unknown stable transfer function parameter. A transfer function is said to be coprime if there are no common pole-zero cancellations. This transfer function parameter must be selected in such a manner that when substituted in Eq. B.5, the desired compensator would be obtained.

In continuation, define the needed transfer functions as follows:

$$N(j\omega) = \frac{a'_1 + b'_1 j}{a'_2 + b'_2 j} \tag{B.6*}$$

---

*Source: (Mousavi Firdeh et al. 2006) © Elsevier

$$D(j\omega) = \frac{c_1 + d_1 j}{c_2 + d_2 j} \tag{B.7*}$$

$$r(j\omega) = \frac{\sum\limits_{i=0}^{i=m} a_i.(j\omega)^i}{1 + \sum\limits_{i=1}^{i=n} b_i.(j\omega)^i} \tag{B.8*}$$

$$P(j\omega) = \frac{e_1 + f_1 j}{e_2 + f_2 j} \tag{B.9*}$$

$$h_{y,u}^D = \frac{N_1 + N_2 j}{P_1 + P_2 j} = \frac{Y_1}{Y_2} \tag{B.10*}$$

where N, D, r, and P are defined as before and $h_{y,u}^D$ is the desired reference linear model. The input–output map in terms of the coprime factors is of the following form:

$$h_{y,u} = -N(s)D(s)r(s) + N(s)P(s) = \frac{Z_1}{Z_2} \tag{B.11*}$$

Then, the exact model matching objective function is formed:

$$J' = \int_{\omega_1}^{\omega_2} \left| \frac{Z_1}{Z_2} - \frac{Y_1}{Y_2} \right|^2 d\omega \tag{B.12*}$$

Minimizing the above function, with respect to the compensator parameters, results in a set of nonlinear algebraic equations that must be solved numerically. However, the goal is to develop a closed-form solution. Notice that under optimality conditions $\frac{Z_1}{Z_2} = \frac{Y_1}{Y_2} \Rightarrow Z_1 Y_2 - Z_2 Y_1 = 0$. Therefore, assuming that a solution exists, the following alternate objective function may be selected:

$$J = \int_{\omega_1}^{\omega_2} |Z_1 Y_2 - Z_2 Y_1|^2 d\omega \tag{B.13*}$$

By substituting Eqs. B.11 and B.6–B.10 into Eq. B.13, the following relation is obtained:

$$E' = \int_{\omega_1}^{\omega_2} \begin{bmatrix} (-AA_1 + BA_2 + B_1 C - B_2 D)^2 + \\ (-AA_2 - BA_1 + B_1 D + B_2 C)^2 \end{bmatrix} d\omega \tag{B.14*}$$

*Source: (Mousavi Firdeh et al. 2006) © Elsevier

where

$$A_1 = a_0 - a_2\omega^2 + a_4\omega^4 - \cdots \quad \text{(B.15*)}$$

$$A_2 = a_1\omega - a_3\omega^3 + a_5\omega^5 - \cdots \quad \text{(B.16*)}$$

$$B_1 = 1 - b_2\omega^2 + b_4\omega^4 - \cdots \quad \text{(B.17*)}$$

$$A = \left(a'_1c_1 - b'_1d_1\right)\left(e_2P_1 - f_2P_2\right) - \left(a'_1d_1 + b'_1c_1\right)\left(f_2P_1 + e_2P_2\right) \quad \text{(B.18*)}$$

$$B = \left(a'_1c_1 - b'_1d_1\right)\left(f_2P_1 + e_2P_2\right) + \left(a'_1d_1 + b'_1c_1\right)\left(e_2P_1 - f_2P_2\right) \quad \text{(B.19*)}$$

$$C = E - G \quad \text{(B.20*)}$$

$$D = F - H \quad \text{(B.21*)}$$

$$E = \left(a'_1e_1 - b'_1f_1\right)\left(c_2P_1 - d_2P_2\right) - \left(a'_1f_1 + b'_1e_1\right)\left(c_2P_2 + P_1d_2\right) \quad \text{(B.22*)}$$

$$F = \left(a'_1e_1 - b'_1f_1\right)\left(c_2P_2 + P_1d_2\right) + \left(a'_1f_1 + b'_1e_1\right)\left(c_2P_1 - d_2P_2\right) \quad \text{(B.23*)}$$

$$G = \left(a'_2c_2 - b'_2d_2\right)\left(e_2N_1 - f_2N_2\right) - \left(a'_2d_2 + b'_2c_2\right)\left(e_2N_2 + f_2N_1\right) \quad \text{(B.24*)}$$

$$H = \left(a'_2c_2 - b'_2d_2\right)\left(e_2N_2 + f_2N_1\right) + \left(a'_2d_2 + b'_2c_2\right)\left(c_2N_1 - f_2N_2\right) \quad \text{(B.25*)}$$

By imposing the necessary optimality condition, i.e., $\nabla_{a,b} = 0$, a set of simultaneous linear algebraic equations is obtained as follows:

$$XY = Z \quad \text{(B.26*)}$$

where

$$X = \begin{bmatrix} T_0 & 0 & -T_2 & \cdots & R_1 & S_2 & -R_3 & \cdots \\ 0 & T_2 & 0 & \cdots & -S_2 & R_3 & S_4 & \cdots \\ -T_2 & 0 & T_4 & \cdots & -R_3 & -S_4 & R_5 & \cdots \\ \vdots & \vdots & \vdots & \vdots & \vdots & \vdots & \vdots & \vdots \\ R_1 & -S_2 & -R_3 & \cdots & Q_2 & 0 & -Q_4 & \cdots \\ S_2 & R_3 & -S_4 & \cdots & 0 & Q_4 & 0 & \cdots \\ -R_3 & S_4 & R_5 & \cdots & -Q_4 & 0 & Q_8 & \cdots \\ \vdots & \vdots & \vdots & \vdots & \vdots & \vdots & \vdots & \vdots \end{bmatrix} \quad \text{(B.27*)}$$

*Source: (Mousavi Firdeh et al. 2006) © Elsevier

$$Y = \left[a_0 \; a_1 \; a_2 \cdots b_1 \; b_2 \; b_3 \cdots \right]^T \tag{B.28*}$$

$$Z = \left[R_1 \; -S_2 \; -R_3 \cdots 0 \; Q_2 \; 0 \cdots \right]^T \tag{B.29*}$$

$$T_l = \int_{\omega_1}^{\omega_2} \omega^l (A^2 + B^2) \, d\omega \tag{B.30*}$$

$$R_l = \int_{\omega_1}^{\omega_2} \omega^l (AD - BC) \, d\omega \tag{B.31*}$$

$$S_l = \int_{\omega_1}^{\omega_2} \omega^l (AC + BD) \, d\omega \tag{B.32*}$$

$$Q_l = \int_{\omega_1}^{\omega_2} \omega^l (C^2 + D^2) \, d\omega \tag{B.33*}$$

Therefore, by supplying the linear plant, the desired closed-loop transfer function, and the frequency range of interest, one could use Eq. B.26 and directly solve for the unknown function parameter $r(s)$.

*B.1.1 Discussion (Nassirharand and Karimi 2004c)***

1. "Desired frequency range" – Based on the experience of the authors, the frequency range of interest is not of major concern. The user may determine the bandwidth of the desired closed-loop system, and select the lower and upper bounds of the desired frequency range. Then, the developed approach may be executed; if the transient response is not satisfactory, then the upper and/or the lower frequency limits may be multiplied by an arbitrary factor such as 2^α (where α is a trial number) until the desired transient response is obtained. Similarly, if the steady-state response is not satisfactory, then the lower and/or the upper limits may be divided by an arbitrary factor such as 2^α until the desired steady-state response is obtained. The procedure is not very sensitive to the selected frequency range. Based on application of the software to a number of different problems, it is recommended that $w1 = 0$ and $w2 = 10$ are to be used as a starting point; this recommended range may have to be modified as outlined earlier.

---

*Source: (Mousavi Firdeh et al. 2006) © Elsevier

**Source: (Nassirharand and Karimi 2004c) © Manchester University Press

2. "Bad frequency range" – If a bad frequency range is used, the developed procedure will not find the desired solution. The question is when the user can tell if a bad frequency range is used? The user may be assured that a bad frequency range is not used if he has gone through the exercise of item 1 above without any useful results. In this case, the desired performance measures may have to be modified or a different feedback structure may have to be selected.
3. "Desired transfer function" – For most problems the desired closed-loop transfer function would be a second order transfer function that possesses the desired dominant poles, and satisfies the desired steady-state error conditions. The user may translate the time and/or frequency performance measures to the desired natural frequency and the desired damping ratio by considering the performance specification equations in standard automatic control textbooks (Rowland 1986); then the steady-state error specifications determine if a zero is also required. The zero may be determined from the relations for definition of a steady-state error. Also, in some cases, additional high order dynamic terms may have to be introduced to balance the matching of polynomial coefficients (see Eq. B.13); this is called enhancing the desired transfer function (Pujara 1988).
4. "Meaningless designs" – If the designed parameters of the controller are such that results in an unstable closed-loop feedback system, then the designer must explore the possibility of modifying the required performance measures or a different feedback structure may have to be assumed.
5. "Improper designs" – It is possible that the model-matching solution is an improper transfer function; as is known, improper transfer functions are not realizable. In such cases, one approach to accommodate such designs is to include adequate high-order dynamic terms until the order of the denominator polynomial equals that of the numerator. Alternatively, one could add εs^n terms ($n = 1, 2, ...$) to the denominator until the order of the denominator equals that of the numerator. This is a standard practice in realizing such improper transfer functions as a pure derivative. The numerical value of ε must be selected such that this does not require unreasonably small integration step size, and it also must not be so large that it alters the behavior of the original transfer function.
6. "Pole-zero shifting" – It should be noted that as the order of the plant increases, the traditional techniques, which give insight into how the pole-zero locations of the controller affect the response, lose effectiveness. As was mentioned earlier, the presented approach complements the traditional techniques (such as pole-zero shifting) for design of classical controllers; for simple plants, the developed tool may be used to verify results, and for complicated plants, where application of the pole-zero shifting approach becomes difficult, the tool may be used to arrive at a solution which may then be analyzed in order to determine where the poles and zeros of the controller ought to be placed for optimum results.

B.1.2 Systematizing the Design Approach

In order to complete the design procedure above, the user must take into account the remarks made in the above Discussion section. This results in a trial-and-error scheme by which the determination of the desired frequency range, determination of the enhancing requirements, and determination of the orders of numerator and denominator of the transfer function parameter (in case of n-th order linear controller design) would be possible. This trial-and-error scheme becomes tedious and time consuming, and at the end the user is not sure if all the possibilities have been considered.

In order to alleviate this problem, the following algorithm is developed:

Algorithm B.1

1. Let the user set the lower and upper limits of desired frequency range of interest; the default values are 0 and 10.
2. Solve design equation (e.g., Eq. B.26) for controller parameters.
3. Validate design.
4. If satisfactory results are obtained, then stop; otherwise, vary the lower and upper limits of the desired frequency range of interest using numerical optimization.
5. If the variation of lower and upper limits of integration is exhausted with no solution, then go to Step 6; otherwise go to Step 2.
6. If an n-th order controller is being designed, increase the degrees of numerator and denominator of the function parameter, and go to Step 2. If increasing the degrees of numerator and denominator of the function parameter is exhausted with no solution, then go to Step 7; otherwise stop.
7. Provide plant transfer function enhancing (by introducing high-order dynamic terms to increase the order of the desired transfer function), and go to Step 2. If enhancing does not produce the desired result, conclude that a solution does not exist; consider relaxing the performance measures or consider a different feedback structure.

B.2 Software

A MATLAB® function based on the above versatile controller design approach is available (Nassirharand 2009b). This function allows design of lead-lag, PID, proportional plus rate feedback, proportional-integral plus rate feedback, and n-th order linear compensators. The primary inputs to this are the plant transfer function and the desired closed-loop behavior in terms of an n-th order transfer function.

The output is the desired controller parameters. The unified design approach that is implemented is based on minimizing a mean-squared error type objective function as given by Eq. B.13. This minimization gives the controller parameters in terms of a set of linear simultaneous algebraic equations. This solution method requires the user to input such parameters as the lower and upper limits of frequency range of interest as well as those discussed in Sect. B.1.1; such parameters could be automatically determined by the use of numerical optimization as outlined in Algorithm B.1.

A MATLAB® function is developed that automates the design of classical controllers. This function is of the following form:

$$gc = cad\_controller\,(gp, hd, params)$$

where

gp is the transfer function of the plant that must be controlled
hd is the transfer function of the desired closed-loop system
params is an array of size 6 that controls the execution of the function

params(1): this input is between numbers 1 and 5. A 1 implies design of a lead-lag compensator, a 2 implies design of a PID compensator, a 3 implies design of an n-th order compensator using the factorization approach, a 4 implies design of a proportional and rate-feedback compensator, and a 5 implies design of a proportional-integral plus rate-feedback compensator.
params(2): the final time to be used for simulation of the final feedback system; if set equal to -1, then the final simulation time value is automatically determined from MATLAB®.
params(3): the desired error between step response of *hd* and the designed closed-loop system; if set equal to -1, then a value of 1 is used.
params(4): the integration step size; if set equal to -1, then a value of 0.005 is used.
params(5): the lower limit of the frequency range of interest; if set equal to -1, then a value of 0.01 rad/s is used.
params(6): the upper limit of the frequency range of interest; if set equal to -1, then a value of 10 rad/s is used. The maximum allowed value is arbitrarily set to 20 rad/s.

gc is the output of the function. In case of lead-lag, PID, and *n*th order compensators, *gc* is the transfer function of the compensator; in case of proportional plus rate-feedback design, *gc* is an array whose first element is the gain of the rate-feedback term, and its second element is the gain of the proportional term; in case of proportional-integral plus rate-feedback, *gc* is an array whose first element is the gain of the integral action, its second element is the gain of the proportional term, and its third element is the gain of the rate-feedback term.

Appendix C
Additional Example Problems for Multivariable Single-Range Controller Design

C.1 Example Problem: One Combustion Chamber Liquid Propellant Engine

The design Algorithm 3.1 is applied to a problem of the sort encountered in propulsion control engineering. The specific propulsion system is a liquid propellant engine of the same class whose simplified schematic drawing is shown in Fig. 3.1. A brief engine operation description along with sample propulsion equations of motion is given in Chap. 3. The primary difference between the present process and that studied in Chap. 3 is due to specific constants and input data which make this a coupled nonlinear multivariable system.

The nonlinear and dynamic computer model of this liquid propellant engine is given, and a controller must be designed to control the combustion chamber pressure; it must also regulate the difference in exit pressure of the oxidizer shut-down valve and that of the stabilizer in order to control the mixture ratio. In short, this plant has two actuating signals (the regulator valve piston travel and the stabilizer valve piston travel). It also has two outputs (the combustion chamber pressure and the difference between the pressure at the exit of the shut-down valve (not shown) and the exit pressure of the stabilizer valve. The shut-down valve is in the oxidizer path, and it is placed between the oxidizer pump and the combustion chamber. Also, it is required that speed of response of both outputs to be as fast as possible without any overshoot. Algorithm 3.1 is executed as follows (Nassirharand and Karimi 2004a).

Step 1 – Trial and error is used, and the following multivariable plant is defined as the reference linear model to be used during the design process.

$$G_{ref} = \begin{bmatrix} \dfrac{25^2}{s^2+50s+25^2} & 0 \\ 0 & \dfrac{25^2}{s^2+50s+25^2} \end{bmatrix} \tag{C.1*}$$

*Source: (Nassirharand and Karimi 2004a) © Sharif University of Technology

A. Nassirharand, *Computer-aided Nonlinear Control System Design: Using Describing Function Models*, DOI 10.1007/978-1-4471-2149-7,
© Springer-Verlag London Limited 2012

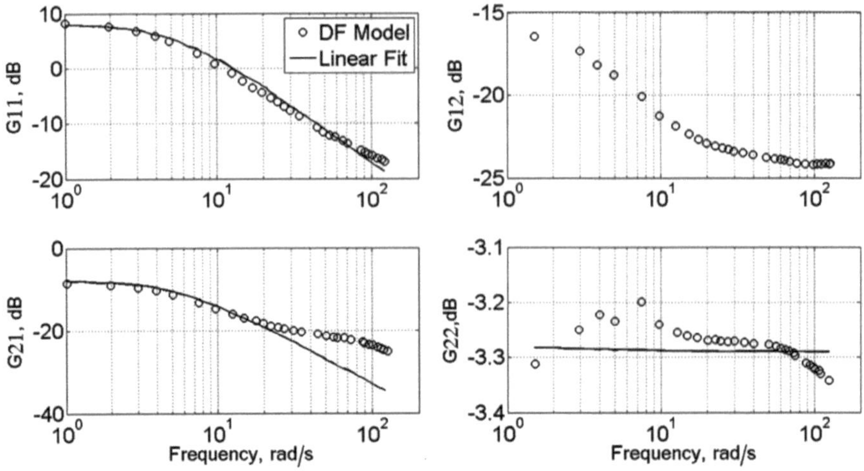

Fig. C.1 Matrix magnitude plot of pseudofrequency response

The above reference linear model is chosen so the closed-loop system would be decoupled, and the diagonal dynamics correspond to a linear second-order system whose natural frequency and damping ratio are 25 and 1, respectively. Use of a higher value for the natural frequency would result in nonlinear system response to have an overshoot, and the use of a lower value for the damping ratio would result in nonlinear system response to be very sluggish.

Step 2 – For the nominal operating regime of interest, the input/output behavior of the process is characterized as outlined in Chap. 2. The nominal value of the regulator control valve piston position and the nominal value of the stabilizer control valve piston position (along with their limits) determine the DC part of excitation, and the maximum travel values of these pistons are used to determine the amplitudes of excitation. From the knowledge of the physical process, the frequency range of interest is determined to be between 1 and 150 rad/s; as was mentioned, the selected frequency pair must not be identical, and they are chosen as nearly equal and related rationally. Initially, the phase shifts are set equal to zero; if satisfactory results are not obtained, then effects of phase shift may also be considered. In continuation, the plant is excited and the outputs (combustion chamber pressure and the difference in shut-down valve exit pressure and the stabilizer control valve exit pressure) are determined. Then, Fourier integrals are evaluated when the outputs are at steady state. Finally, the matrix SIDF model is calculated. The resulting matrix SIDF model and the fitted linear model of the next step are depicted in Figs. C.1 and C.2.

Step 3 – In this step, by providing the real and imaginary parts of the pseudofrequency response obtained in the previous step as a function of frequency, the *invfreqs* function of the MATLAB® software is used to identify the individual elements of the matrix linear approximation one at a time. This function allows

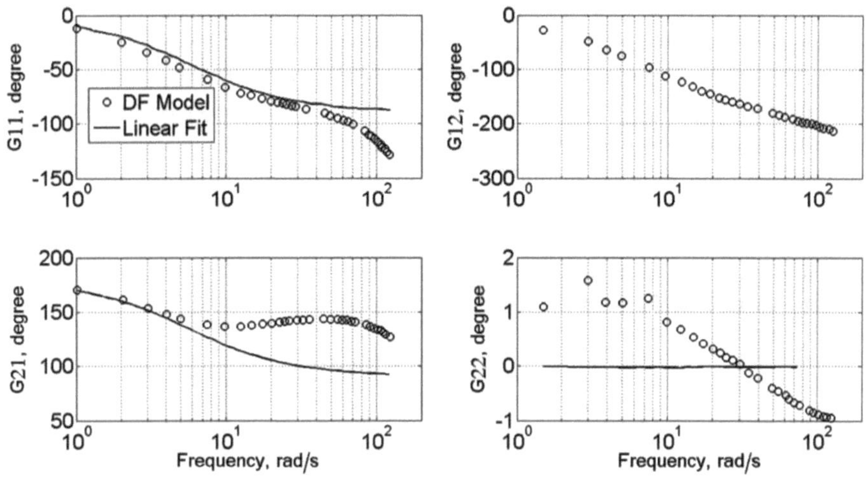

Fig. C.2 Matrix phase plot of pseudofrequency response

introduction of weights to weigh the data at different frequency ranges differently. In this example, no weights are required to obtain a reasonable fit. It should also be noted that the numerator and denominator degrees of the transfer function polynomials are also an input that is determined by trial and error. For example, the degrees of the numerator and denominator polynomials of the transfer function to be identified are set to 1 initially, and they are increased until no improvement in the quality of fit is observed. By examining Fig. C.1, it is apparent that the contribution of g_{12} (where g_{12} corresponds to the first row and second column of the plant matrix SIDF model) is very small when compared to contributions of the g_{11} element, and it may be set equal to zero.

The identified linear model is of the following form:

$$G(s) = \begin{bmatrix} \frac{13.99}{s + 5.617} & 0 \\ \frac{-2.254}{s + 5.617} & \frac{0.6848s + 3.8493}{s + 5.617} \approx 0.6853 \end{bmatrix} \quad \text{(C.2*)}$$

Step 3 – The diagonal elements of the controller are determined by solving for $r(s)$ using Eq. B.26 followed by application of Eq. B.5. Determination of $c_{11}(s)$ is described, and determination of $c_{22}(s)$ is similar. In order to apply Eq. B.26, the co-prime factor pairs ($[N(s), D(s)]$ and $[P(s), Q(s)]$) must be determined. Since $g_{11}(s)$ is stable, one could choose these factors as follows: $N(s) = g_{11}(s), D(s) = 1$, $P(s) = 0$, and $Q(s) = 1$. From Eq. B.1 it is concluded that $h^D_{y,u} = \frac{25^2}{s^2+50s+25^2}$. The only remaining parameters that need to be specified in order to solve for $r(s)$

*Source: (Nassirharand and Karimi 2004a) © Sharif University of Technology are ω_1, ω_2 in Eqs. B.30–B.33 and m, n in Eq. B.8. Some trial and error must be employed to determine these parameters; the procedure is computer aided, and therefore, the cycle time for determination of these parameters is not very long. For starting point, set $m = n = 0$ and $\omega_1 = 0, \omega_2 = \omega_{BW}$, where ω_{BW} is the bandwidth. To improve on the quality of behavior at high frequencies, increase ω_2 and/or increase ω_1, and in order to improve the quality of behavior at low frequencies, decrease ω_2. If satisfactory results are not obtained, then increase m, n. Repeat this cycle until a design is obtained. For this example, $\omega_1 = 0, \omega_2 = 10, m = 2,$ *and* $n = 2$. Then Eq. B.26 is applied, and $r(s)$ is determined. Upon substitution of $r(s)$ in Eq. B.5, the desired $c_{11}(s)$ is obtained as follows:

$$c_{11}(s) = \frac{44.67s + 250.9}{s^2 + 50s} \tag{C.3*}$$

The off-diagonal elements are determined from application of Eq. B.2:

$$c_{12}(s) = 0, c_{21}(s) = \frac{-146.9s - 824.9}{s^3 + 55.62s^2 + 280.9s} \tag{C.4*}$$

Hence, the synthesized controller is of the following form:

$$C(s) = \begin{bmatrix} \dfrac{44.67s + 250.9}{s^2 + 50s} & 0 \\ \dfrac{-146.9s - 824.9}{s^3 + 55.62s^2 + 280.9s} & \dfrac{0.01979s + 911.3}{s^2 + 49.91s} \end{bmatrix} \tag{C.5*}$$

For comparison purposes, this step is also solved by application of a H_∞ design procedure (Safonov et al. 1989) that is also automated by MATLAB® function *hinf*. The considered feedback structure is shown in Fig. C.3. In this figure, R_1 is the set point, Z_1 is the weighted control signal, Z_2 is the weighted disturbance output signal, Z_3 is the weighted error signal, U is the control signal, *Error* is the error signal, and W is a weighting function which may be subscripted for clarity. The following remarks are in order (Nassirharand and Mousavi Firdeh 2009c):

1. In order to compensate for erosion and high levels of friction in mechanical parts, the following control signal weighting function is selected:

$$W_{cntrl} = \begin{bmatrix} \dfrac{0.2\,(s + 40)}{s + 200} & 0 \\ 0 & \dfrac{0.2\,(s + 40)}{s + 120} \end{bmatrix} \tag{C.6**}$$

*Source: (Nassirharand and Karimi 2004a) © Sharif University of Technology

**Source: (Nassirharand and Mousavi Firdeh 2009a) © Elsevier

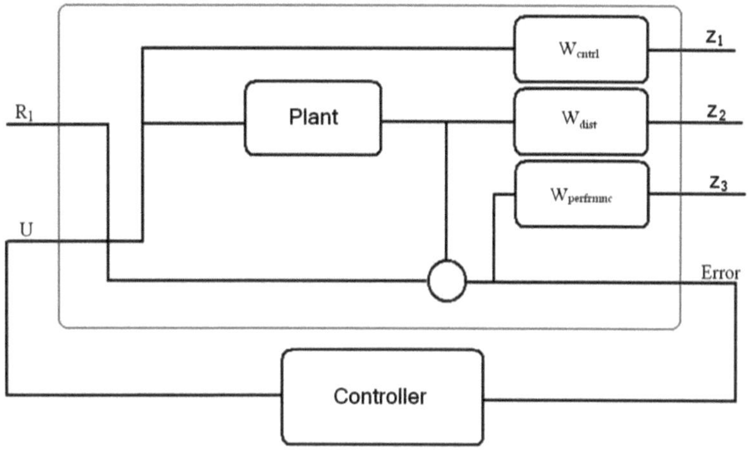

Fig. C.3 Structure of the H_∞ feedback system. Source: (Nassirharand and Mousavi Firdeh 2009a) © Elsevier

The factor 0.2 is a scalar weight, the zero at 40 is selected so the signals with frequencies less than the required bandwidth would not be penalized, and the poles at 200 and 120 are used to severely penalize the control signals corresponding to the two actuating signals with frequencies higher than 200 and 120 rad/s, respectively.

2. Considering a multiplicative uncertainty, the following disturbance weighting function is selected to achieve robust stability:

$$W_{dist} = \begin{bmatrix} \dfrac{0.2\,(s+1)}{s+20} & 0 \\ 0 & \dfrac{0.2\,(s+1)}{s+20} \end{bmatrix} \tag{C.7*}$$

This choice reflects a 5% uncertainty for low frequencies (here taken to be 1 rad/s) and a 100% uncertainty for higher frequencies (here taken to be 20 rad/s).

3. Since the natural frequency is 25 rad/s and since it is desired to have a small steady-state error, the following weight for W_{cntrl} is selected:

$$W_{prfrmnc} = \begin{bmatrix} \dfrac{0.2\,(s+30)}{s+0.001} & 0 \\ 0 & \dfrac{0.2\,(s+30)}{s+0.001} \end{bmatrix} \tag{C.8*}$$

*Source: (Nassirharand and Mousavi Firdeh 2009a) © Elsevier

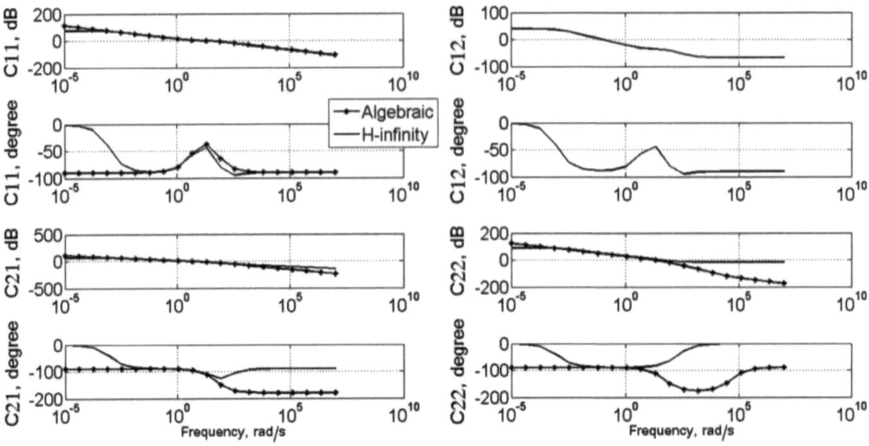

Fig. C.4 Bode plots of the algebraic and H_∞ controllers. Source: (Nassirharand and Mousavi Firdeh 2009a) © Elsevier

A reduced order model of the H_∞ controller, which is designed using the mentioned MATLAB® function, follows:

$$C_\infty(s) = \begin{bmatrix} \dfrac{26.91s^2 + 5536s + 3.025e4}{s^3 + 161s^2 + 5235s + 5.235} & \dfrac{0.0004415s^3 - 0.5571s^2 - 125.7s - 685.5}{s^3 + 170s^2 + 7018s + 7.018} \\ \dfrac{0.7272s + 109.5}{s^2 + 38.67s + 0.03867} & \dfrac{0.1711s + 34}{s + 0.001} \end{bmatrix} \tag{C.9*}$$

The frequency responses of the two designed controllers are compared in Fig. C.4. The following observations are made:

1. Although the gain of the (1,1) element of C_∞ is slightly higher than that of the algebraic controller, the dynamic behavior of the (1,1) elements is similar. It is clear that the phases of (1,1) elements are almost identical.
2. The gain of the (1,2) element of C_∞ rapidly decreases as a function of frequency, and in effect one could set this element to zero similar to the algebraic controller.
3. The gains of the (2,1) elements are almost identical. However, there is a large shift in phase; since the contributions of this element are drastically less than those of the (2,2) element, the effect of difference in the phase shifts on the stabilizer axis is very small. For example, at 2.83 rad/s the gain of the (2,1) element is -0.374 dB while the gain of the (2,2) element is 16.5 dB.
4. The gains and phases of the (2,2) elements up to a frequency of approximately 20 rad/s are identical; beyond this frequency, the gain values are small.

---

*Source: (Nassirharand and Mousavi Firdeh 2009a) © Elsevier

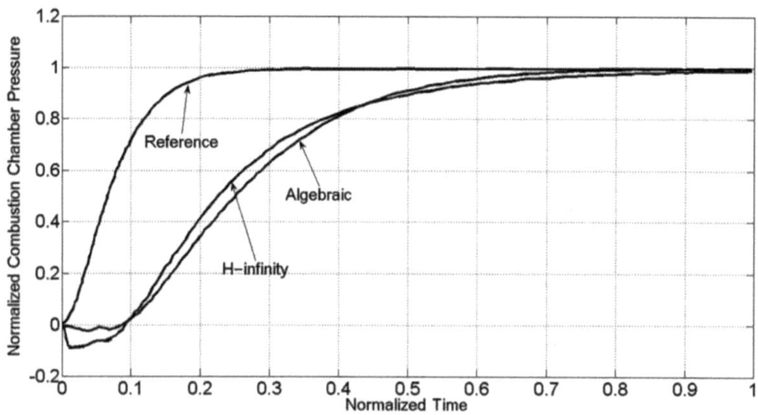

Fig. C.5 Time history of the combustion chamber pressure. Source: (Nassirharand and Mousavi Firdeh 2009a) © Elsevier

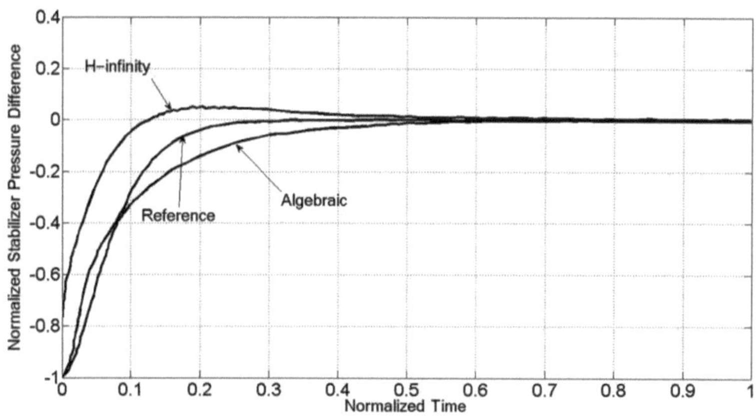

Fig. C.6 Time history of the stabilizer pressure difference. Source: (Nassirharand and Mousavi Firdeh 2009a) © Elsevier

From the above observations, it may be concluded that the dynamic behaviors of the two controllers are similar; however, the algebraic controller is simpler; and hence, it is easier to implement.

Step 4 – Finally, the design is verified. The combustion chamber pressure and the difference in input and output stabilizer pressures for the two designs in cascade with the nonlinear plant along with desired responses are depicted in Figs. C.5 and C.6. The corresponding time histories of the control signals are shown in Figs. C.7 and C.8. It is apparent that control system behavior is acceptable. To check the robustness of the designed control system, the pumps' heads were reduced by 10%. The corresponding time responses are depicted in Figs. C.9 and C.10. Again, it is

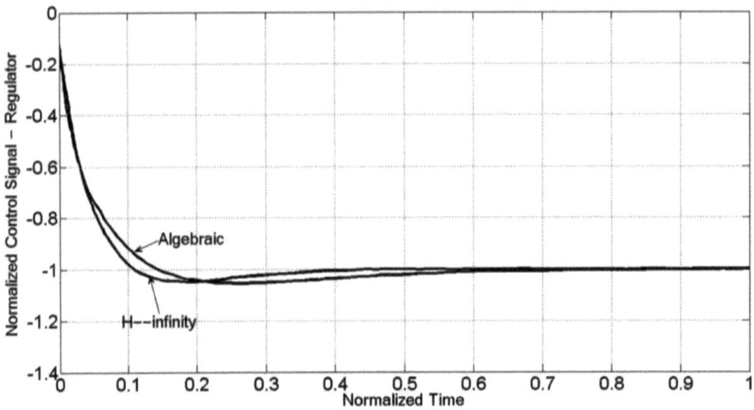

Fig. C.7 Time history of the regulator control signal. Source: (Nassirharand and Mousavi Firdeh 2009a) © Elsevier

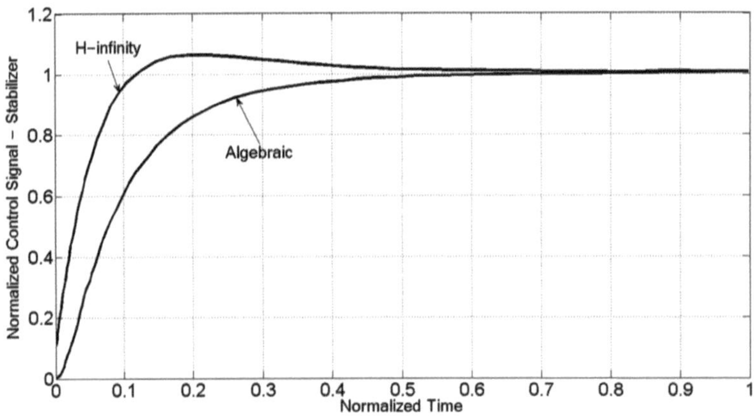

Fig. C.8 Time history of the stabilizer control signal. Source: (Nassirharand and Mousavi Firdeh 2009a) © Elsevier

apparent that system is responding as desired. With reference to Figs. C.5, C.6, C.7, C.8, C.9, and C.10, the following observations are made:

1. The combustion chamber pressure response with the H_∞ controller includes a higher level of undershoot than that of the algebraic controller, which may not be desirable. Qualitatively, the remaining parts of the two responses may be considered to be equal.
2. The stabilizer pressure difference response with the H_∞ controller has a faster rise time than that of the response with the algebraic controller, and it has a small amount of overshoot. However, the settling times may be considered to be identical. There is no overshoot with the algebraic controller.

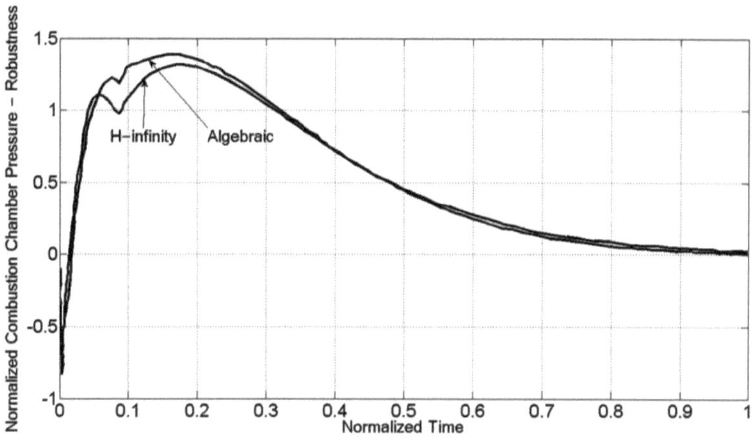

Fig. C.9 Robustness test – combustion chamber pressure. Source: (Nassirharand and Mousavi Firdeh 2009a) © Elsevier

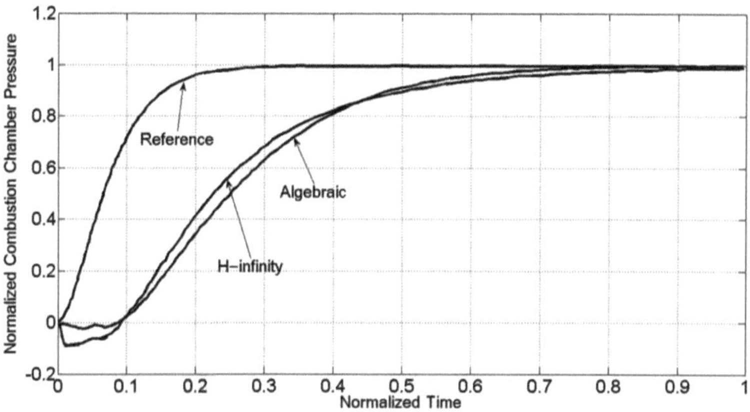

Fig. C.10 Robustness test – stabilizer pressure difference. Source: (Nassirharand and Mousavi Firdeh 2009a) © Elsevier

3. Control signals of both the regulator and stabilizer axis are smooth and acceptable.
4. The time histories of the outputs under a drop in pumps' pressures are competitive.

C.2 Example Problem: Five Combustion Chamber Liquid Propellant Engine

The controller design Algorithm 3.1 is applied to the propulsion control of a five combustion chamber liquid propellant engine (LPE). The schematic drawing of the considered liquid propellant engine is shown in Fig. C.11. A brief engine operation

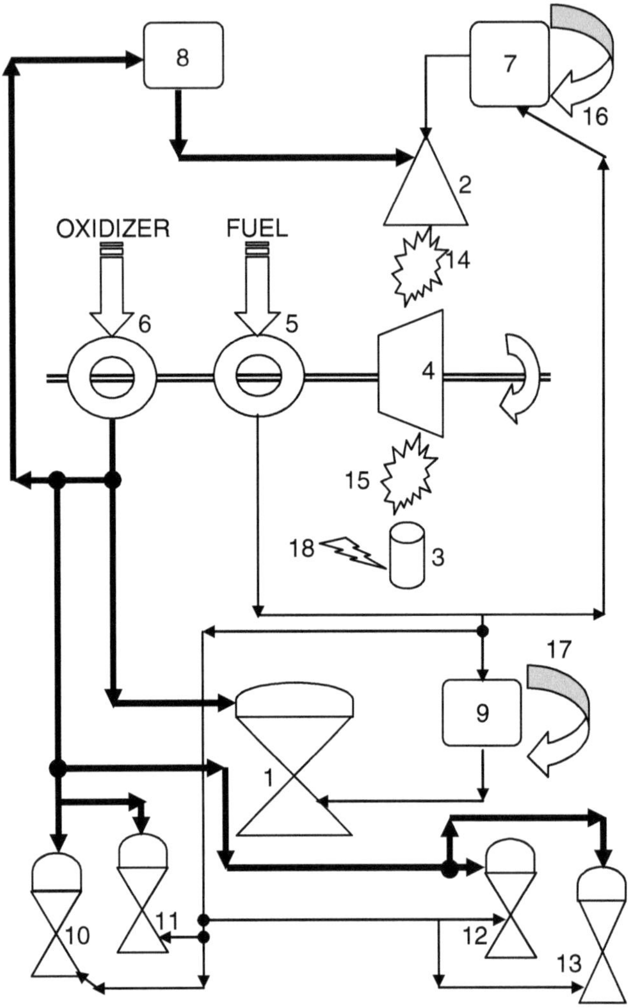

Fig. C.11 Schematic diagram of the LPE (see Table C.1 for key)

description is given next, and the corresponding mathematical model is similar to that presented in Chap. 3.

This engine is composed of the following elements (not all are shown in Fig. C.11): (1) combustion chamber, (2) gas generator, (3) starter, (4) turbine, (5) fuel pump, (6) oxidizer pump, (7) fuel diaphragm valve, (8) stabilizer control valve, (9) oxidizer shut-down valve of the gas generator, (10) oxidizer diaphragm valve, (11) oxidizer shut-down valve of the combustion chamber, (12) regulator control valve, (13) fuel discharge valve, (14) fuel shut-down valve of the gas generator, (15) exhaust, (16)–(19) thrusters (or *verniers*), (20) fuel shut-down valve of thrusters,

Table C.1 Key for Fig. C.11

No.	Description
1	Combustion chamber
2	Gas generator
3	Starter
4	Turbine
5	Fuel pump
6	Oxidizer pump
7	Regulator valve
8	Gas generator stabilizer valve
9	Combustion chamber stabilizer valve
10–13	Thruster
14	Gas generator combustion products
15	Starter combustion products
16, 17	Control signal

(21) oxidizer shut-down valve of the thrusters, (22) staged start valve, (23) valve for simultaneous discharge of the tanks, (24)–(35) orifices, (36)–(49) pipes, and (50) nonreturn valve.

The LPE operates in the following manner. First, the pressures inside the fuel and oxidizer tanks reach to specified values. Then, electric commands to starter, fuel, and oxidizer diaphragm valves (not shown) are issued to start the turbine and allow fuel and oxidizer to enter the system. As the exit pressurized gas of the starter turns the turbine, the shaft of the turbopump starts rotating, and the pumps are commissioned. As a result, fuel and oxidizer would be pumped.

In the fuel path, after liquid modulation, the fuel path is divided into three branches. In the first branch, some fuel enters the stabilizer of the combustion chamber. In the second branch, some fuel is guided toward the regulator, and finally, in the third branch, some fuel heads toward the thrusters. In the first branch, the combustion chamber stabilizer guides a controlled amount of fuel to the fuel discharge valve (not shown). Under the condition that fuel discharge valve is not activated, the fuel that enters this valve travels toward the combustion chamber after going through appropriate connections. In the second branch, the amount of fuel that is guided toward the regulator valve passes through appropriate connections, and it enters the regulator valve. The regulator control valve passes a controlled amount of fuel to the gas generator. Finally, in the third branch, some fuel enters the thrusters.

In the oxidizer path, the oxidizer exiting the oxidizer pump is divided into two branches. In one branch, the oxidizer enters the combustion chamber. This amount of oxidizer and the amount of fuel that enters the combustion chamber from the fuel path are combined, and due to the resulting combustion, the engine is turned on. The second branch itself is divided into two other subbranches or parts.

In the first subbranch, some oxidizer enters the gas-generator stabilizer control valve; this stabilizer sends a controlled amount of oxidizer into the gas generator. This amount of oxidizer and the amount of fuel that enters the gas generator from the fuel path are combined, and combustion is resulted. The exit pressurized gas

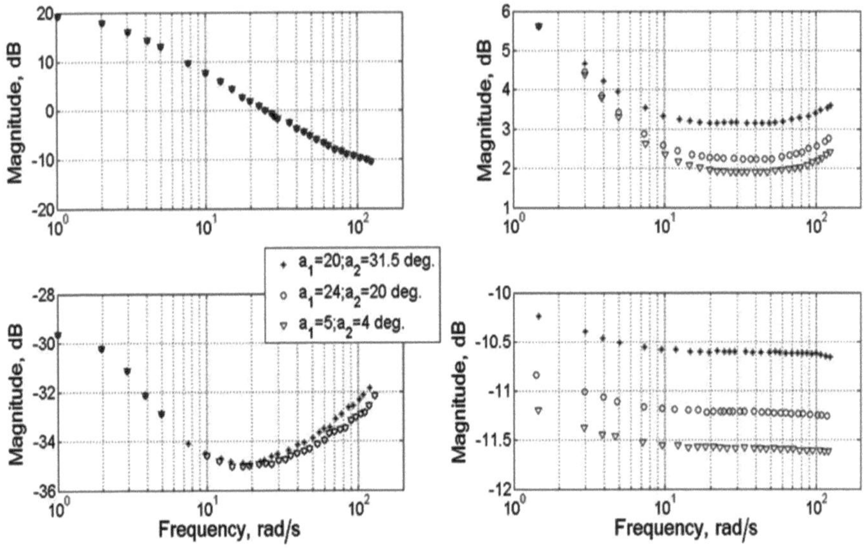

Fig. C.12 Gain plots of SIDF models

of the gas generator keeps the turbine rotating, and the starter gradually leaves the circuit. In the second subbranch, some oxidizer enters the thrusters. This amount of oxidizer that enters the thrusters is combined with the amount of fuel that comes from the fuel path, and combustion takes place.

The nonlinear and dynamic computer model (Karimi and Nassirharand 2006; Nassirharand and Karimi 2004b) of the liquid propellant engine is utilized to design a controller for the system. It is desired that the final nonlinear closed-system input–output behavior would mimic the following reference linear model (Abbassi 2003):

$$G_{ref} = \begin{bmatrix} \dfrac{9.35s + 28.06}{s^2 + 12.15s + 28.06} & 0 \\ 0 & \dfrac{56}{s + 56} \end{bmatrix} \tag{C.10}$$

The controller synthesis procedure steps are executed as described below:

Step 1 – The desired reference linear model is given by Eq. C.10.

Step 2 – The SIDF models of the engine are obtained as outlined in Chap. 2 by simultaneously exciting both input channels with sinusoids of different but nearly equal frequencies, obtaining the output data, and finally postprocessing the output data using the describing function generator software. The corresponding pseudofrequency response plots are depicted in Figs. C.12 and C.13. From the examination of these plots, it is observed that the spread in the SIDF models is not significant. Therefore, it may be concluded that a single-range linear controller might suffice.

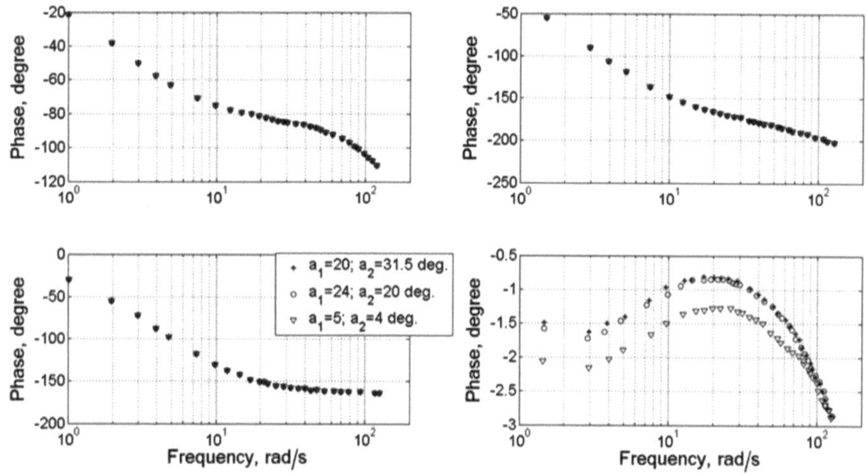

Fig. C.13 Phase plots of SIDF models

Step 3 – In continuation, MATLAB® is used, and the following linear model is identified:

$$G_f(s) = \begin{bmatrix} \dfrac{28.13s + 35850}{s^2 + 1283s + 3534} & \dfrac{4.718s^2 - 1671s + 6923}{s^2 + 1283s + 3534} \\ 0.0 & 0.275 \end{bmatrix} \quad \text{(C.11)}$$

Step 4 – In this step, two different feedback structures are considered for controller design (Abbassi 2003). The first structure is similar to that shown in Fig. B.1 with $g_{21} = 0$; the corresponding feedback structure would be given by Fig. C.14. Substitute appropriate terms into Eq. B.3; then compare the main return ratios for loops 1 and 2 with the main diagonal terms to obtain $c_{11}(s)$ and $c_{22}(s)$. Finally, use Eq. B.2, and solve for $c_{12}(s)$ and $c_{21}(s)$. The designed controller (with the feedback structure given by Fig. C.14) is given below:

$$C(s) = \begin{bmatrix} \dfrac{s+3}{3s} & -\dfrac{8.386s^2 + 2970.14s - 12305.368}{s^2 + 1274.44s} \\ 0 & \dfrac{200}{s} \end{bmatrix} \quad \text{(C.12)}$$

With the second structure (see Fig. C.15), the controller matrix would not be presentable in a standard fashion as in the case with the first structure. With

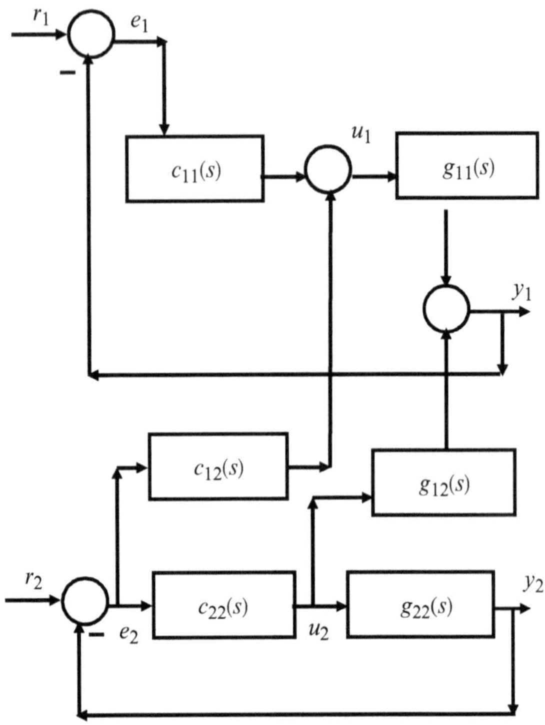

Fig. C.14 Structure of the first considered feedback system

reference to Fig. C.15, c_{11} and c_{22} would be as designed in the first structure. By inspecting the block diagram in Fig. C.15, the following equation is obtained:

$$y_1 = g_{11}c_{11}e_1 + g_{11}Mr_2 + g_{12}u_2 \tag{C.13}$$

To achieve decoupling, the following equation must hold:

$$g_{11}Mr_2 + g_{12}u_2 = 0 \tag{C.14}$$

Note that

$$\frac{u_2}{r_2} = g_{ref}^{22}g_{22}^{-1} = \frac{203}{s + 56} \tag{C.15}$$

where g_{ref}^{22} is the (2,2) element of matrix G_{ref} given by Eq. C.10. Compare Eqs. C.15 and C.14, and substitute for known quantities to obtain M as follows:

$$M = \frac{-33.63s^3 + 11820s^2 - 15970s - 135900}{s^3 + 1339s^2 + 75233s + 189560} \tag{C.16}$$

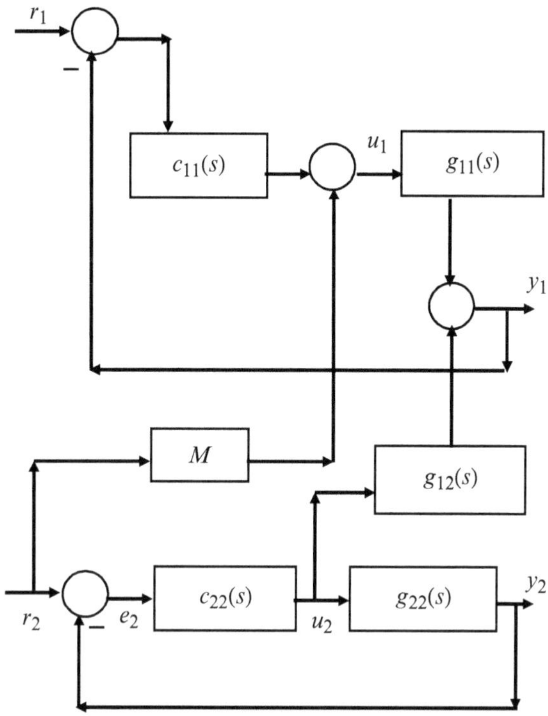

Fig. C.15 Structure of the second considered feedback system

Another controller is designed by minimizing an exact model-matching objective function as follows (Nassirharand and Karimi 2004b):

$$C(s) = \begin{bmatrix} \dfrac{0.4309s^2 + 0.734s + 0.062}{s^2} & -\dfrac{3.6677s^2 + 12.2743s + 52.2724}{0.9248s^3 + 0.2823s^2 + 2.1193s + 0.3302} \\ \dfrac{0.0538s^2 + 0.070044s + 0.1853}{s^2} & \dfrac{75.430s + 0.4989}{s^2 + 0.0133} \end{bmatrix} \tag{C.17}$$

Step 5 – In this step, design is verified by nonlinear simulation. The results of design verification are depicted in Figs. C.16, C.17, C.18, and C.19. In these simulations, step disturbance signals of four units at the inputs of control valves are also included. The depicted simulation results correspond to the two designs developed above and one prior work (Nassirharand and Karimi 2004b). From the examination of the response plots, it is fair to conclude that the second feedback structure would be preferred to the other two designs. Furthermore, unlike the work by Nassirharand and Karimi 2004b, the new designs have disturbance rejection properties. The control signals corresponding to the present designs have a more acceptable behavior when compared to the prior work.

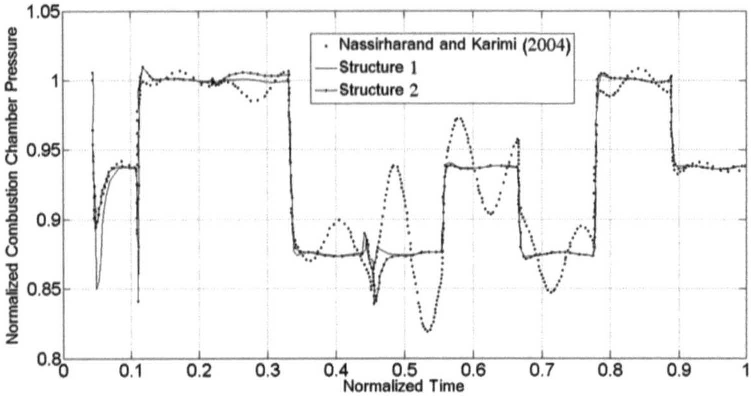

Fig. C.16 Design verification for the combustion chamber pressure (Abbassi 2003)

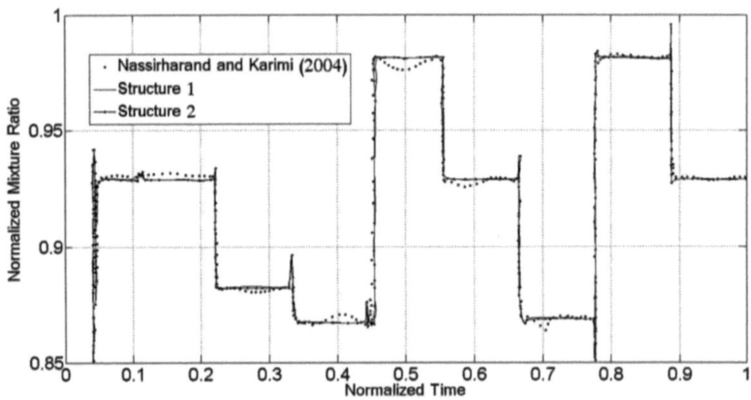

Fig. C.17 Design verification for mixture ratio (Abbassi 2003)

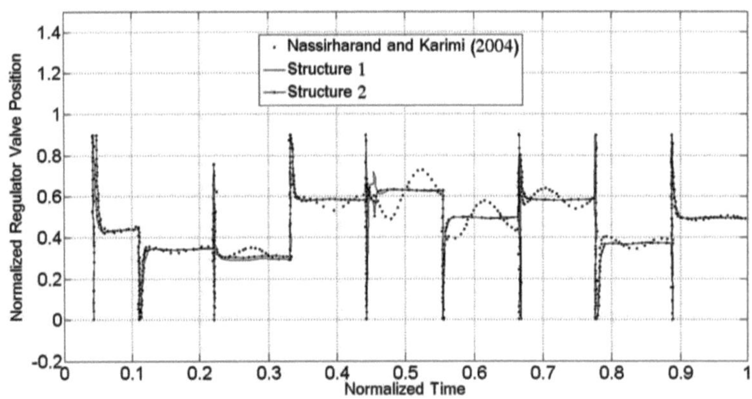

Fig. C.18 Behavior of control signal no. 1 (Abbassi 2003)

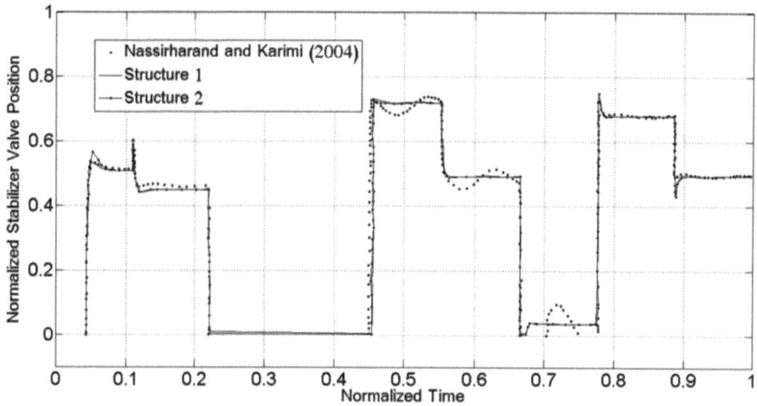

Fig. C.19 Behavior of control signal no. 2 (Abbassi 2003)

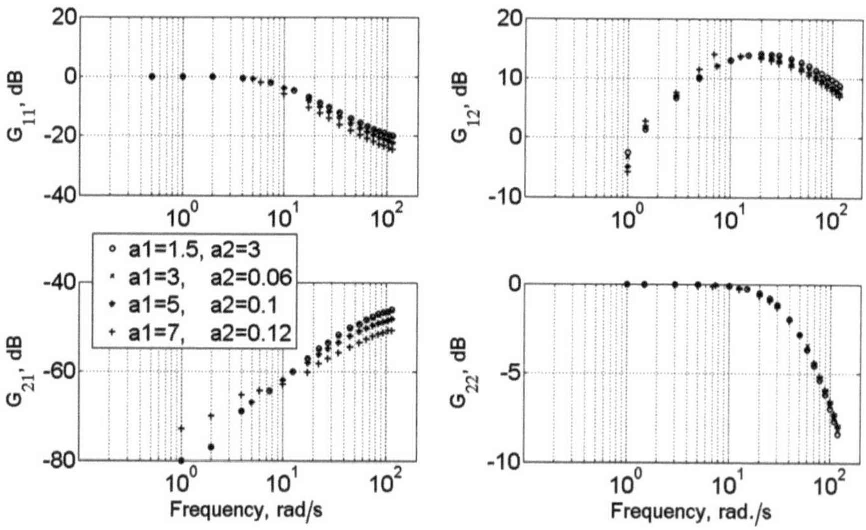

Fig. C.20 Magnitude plot for stability test (structure 1)

As is demonstrated in Figs. C.20 and C.21, the SIDF models of the final closed-loop system at various operating regimes of interest are successfully obtained; hence, one may conclude that closed-loop system is stable when operating at the desired operating regimes.

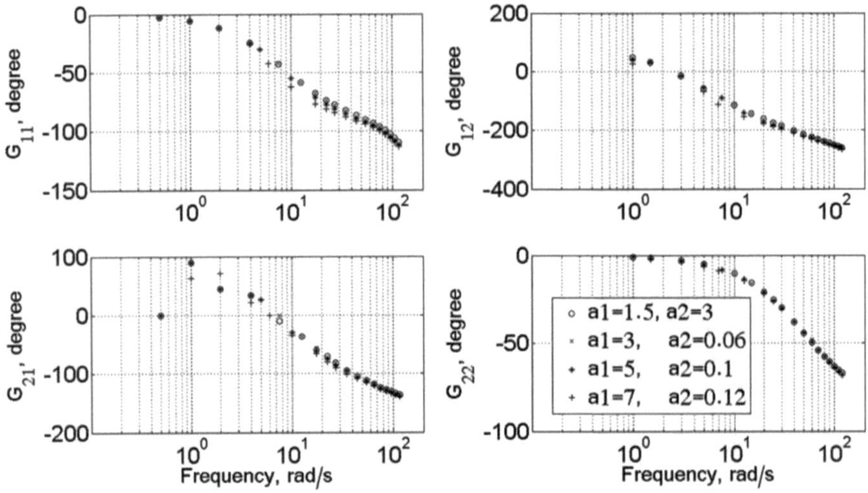

Fig. C.21 Phase plot for stability test (structure 1)

C.3 Example Problem: An Unstable Multivariable Robot Arm

It is emphasized that the basic SRLCD procedure that is outlined in Sect. 3.1 is limited to stable plants. For unstable plants, one approach is to stabilize the plant, and then apply the outlined procedure to the stabilized plant. This approach is demonstrated by the following example problem.

A two degree-of-freedom robot arm is considered (see Fig. C.22; Teh and Nassirharand 2010b). Let L_i, M_i, and q_i be the length, the lumped mass, and angle of link i; for this example problem: $M_1 = 3Kg$, $M_2 = 1Kg$, $L_1 = 1m$, and $L_2 = 1m$ (Utkin et al. 1999). The equations of motion of the robot as given by Utkin et al. 1999 follow:

$$\begin{bmatrix} \tau_1 \\ \tau_2 \end{bmatrix} = \begin{bmatrix} m_{11} & m_{12} \\ m_{21} & m_{22} \end{bmatrix} \begin{bmatrix} \ddot{q}_1 \\ \ddot{q}_2 \end{bmatrix} + \begin{bmatrix} n_1 \\ n_2 \end{bmatrix} \tag{C.18}$$

where

$$m_{22} = L_2^2 M_2 \tag{C.19}$$

$$m_{12} = m_{21} = m_{22} + L_1 L_2 M_2 \cos(q_2) \tag{C.20}$$

$$m_{11} = L_1^2 (M_1 + M_2) + 2m_{12} - m_{22} \tag{C.21}$$

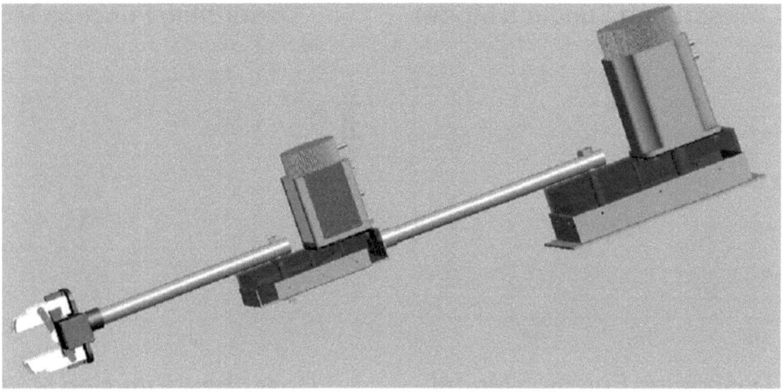

Fig. C.22 A two degree-of-freedom robot arm (Teh 2010)

$$n_2 = L_1 L_2 M_2 \dot{q}_1^2 \sin(q_2) \tag{C.22}$$

$$n_1 = -L_1 L_2 M_2 \left(2\dot{q}_1\dot{q}_2 - \dot{q}_2^2\right) \sin(q_2) \tag{C.23}$$

First, velocity feedback is used to stabilize the system. The gains of the velocity feedback for the first and second links are arbitrarily set to be 50 and 40, respectively. In continuation, Algorithm 3.1 is executed for the stabilized system as follows:

Step 1 – The desired reference linear model is a diagonal matrix transfer function whose (1,1) element is a second-order transfer function with a damping ratio of 0.5 and a natural frequency of 100 rad/s, and its (2,2) element is another second-order transfer function with a damping ratio of 0.5 and a natural frequency of 150 rad/s.

Step 2 – The stabilized system is simulated in the Simulink environment, and the outputs as functions of time are obtained. Then, the MATLAB® software outlined in Chap. 2 is utilized, and the describing function models are generated. These models along with linear system identification results of the next step are shown in Figs. C.23 and C.24 (these figures originally appeared in Nassirharand 2008c).

Step 3 – As was mentioned, the *invfreqs* function of MATLAB® is used, and the following linear model is identified (Govindan 2011):

$$G = \begin{bmatrix} \dfrac{10.78}{s\,(s+7.84)} & \dfrac{-37.99}{s^2+129.8s+667.9} \\ \dfrac{-38.08}{s^2+129.7s+665.6} & \dfrac{-0.003457s+0.0994}{s} \end{bmatrix} \tag{C.24}$$

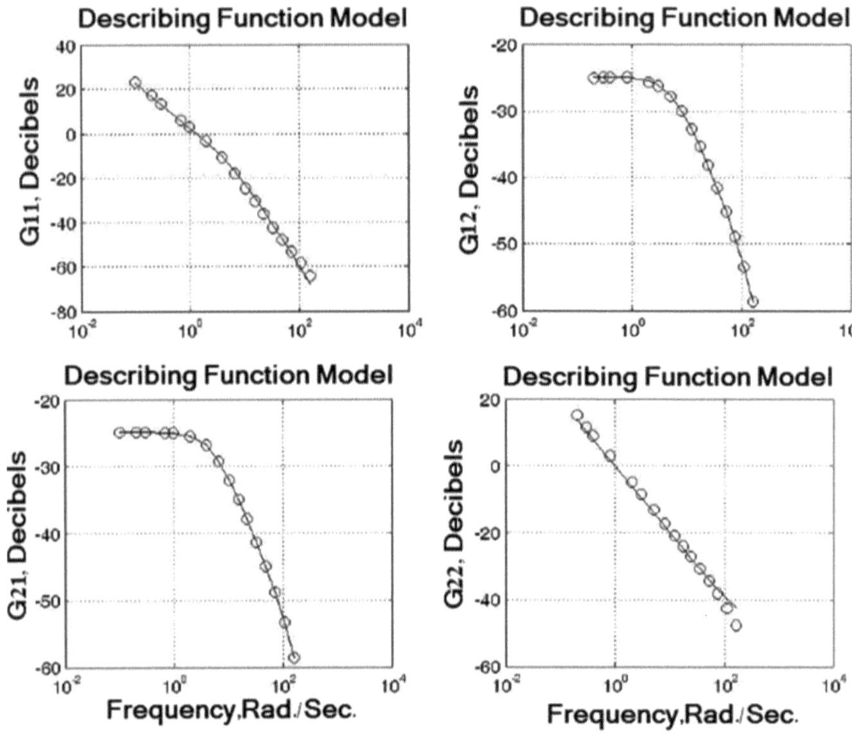

Fig. C.23 The multivariable gain matrix of the robot. Source: (Nassirharand 2008c) © Emeralds

Step 4 – The design equations developed in Appendix B are used to design the controller. For the return ratio corresponding to both loops, the following design parameters are used: $m = n = 6; \omega_1 = 0; \omega_2 = 10$. The designed controller is of the following form (Govindan 2011):

$$C = \begin{bmatrix} \dfrac{2.56}{s^2 + 3.2s + 2.56} & c_{12}(s) \\ c_{21}(s) & \dfrac{4}{s^2 + 4s + 4} \end{bmatrix} \tag{C.25}$$

where

$$c_{12}(s) = \frac{992s^3 + 1.219e5s^2 + 8.988e5s}{s^3 + 289.7s^2 + 3.362e4s + 1.723e5} \tag{C.26}$$

$$c_{21}(s) = \frac{8.526e4s^2 + 5.132e5s}{s^3 + 664.5s^2 + 4.318e4s + 2.213e5} \tag{C.27}$$

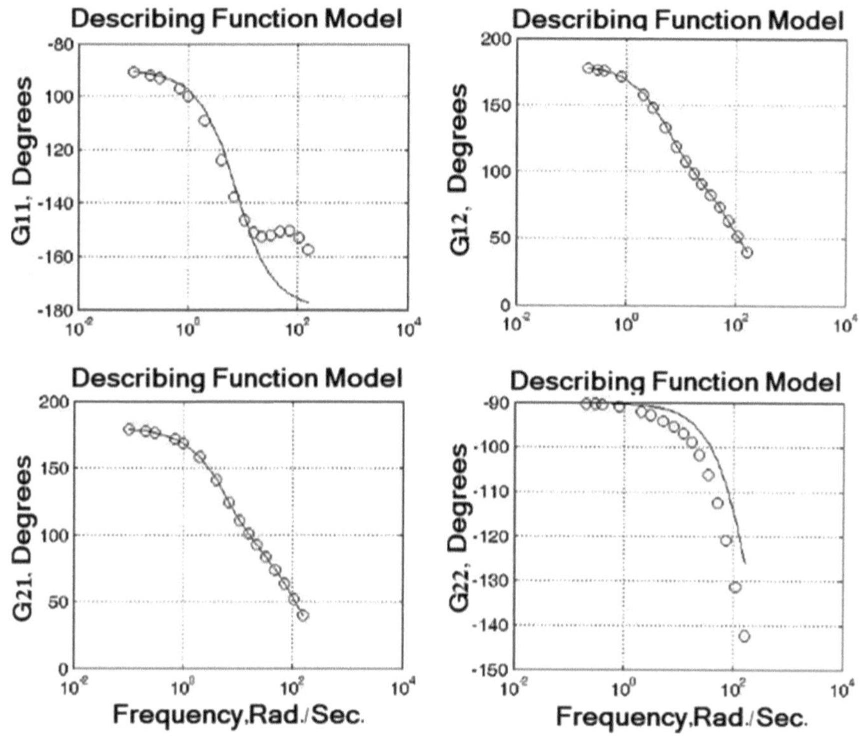

Fig. C.24 The multivariable phase matrix of the robot. Source: (Nassirharand 2008c) © Emeralds

Step 5 – Finally, the design is verified via a nonlinear simulation. The results are compared with an alternate design (Nassirharand 2008c) in Figs. C.25, C.26, C.27, C.28, C.29, and C.30. The following remarks are made:

1. There is less coupling (see Figs. C.26 and C.29) with the new design at the expense of increase in settling time (see Figs. C.25 and C.28).
2. The robustness of both designs is competitive (see Figs. C.25, C.26, C.27, C.28, C.29, and C.30).
3. The control signals of both designs are acceptable. The new design has less control effort (see Figs. C.27 and C.30).

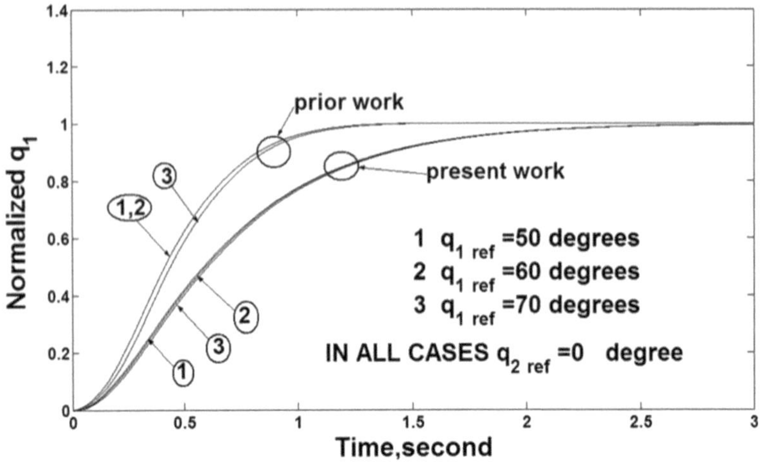

Fig. C.25 Tracking quality of the first axis

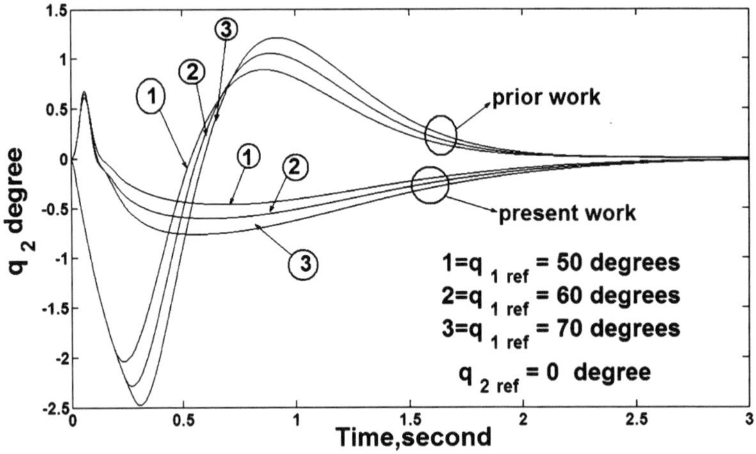

Fig. C.26 Decoupling from the second axis

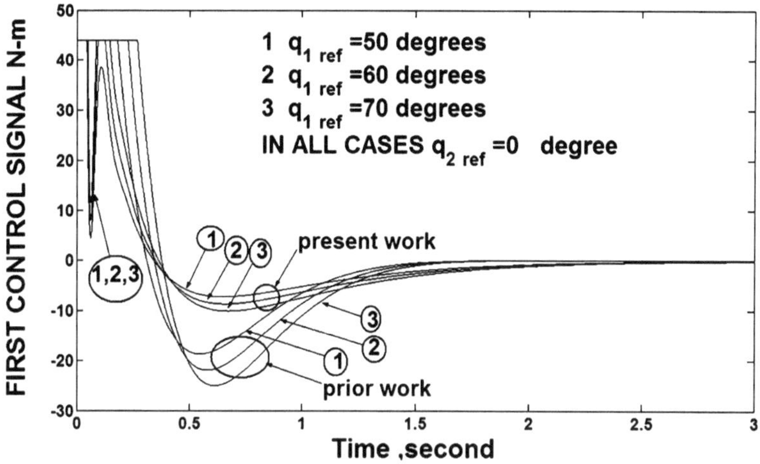

Fig. C.27 Quality of the control signal for the first axis

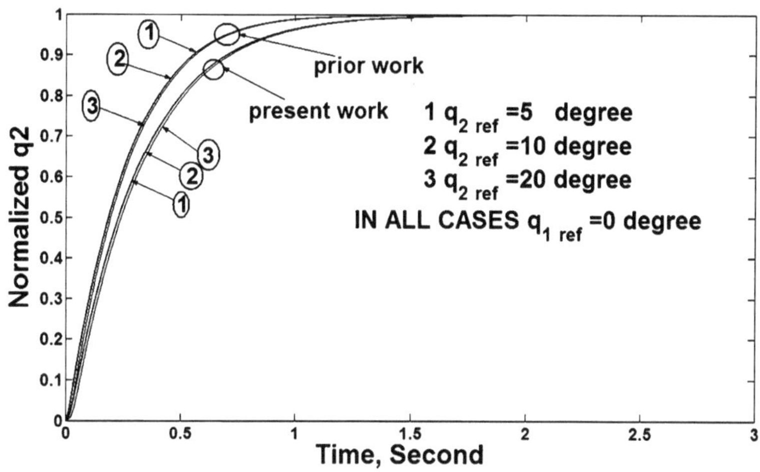

Fig. C.28 Tracking quality of the second axis

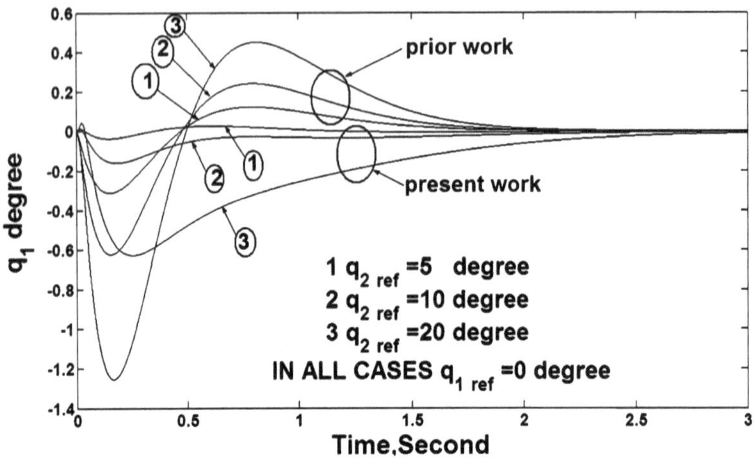

Fig. C.29 Decoupling from the first axis

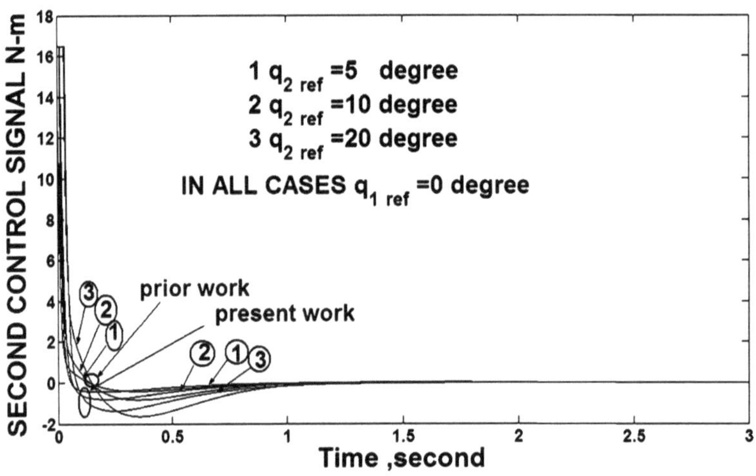

Fig. C.30 Quality of the control signal for the second axis

Appendix D
Dual-Range Controllers and Simultaneous Stabilization Theory

D.1 Bézout's Identity

Application of the factorization theory of linear systems requires one to solve for the coprime factors satisfying the Bezout's identity, i.e.:

$$PN + QD = I \tag{D.1}$$

$$\overline{N}\,\overline{P} + \overline{D}\,\overline{Q} = I \tag{D.2}$$

where

D = a right coprime factor of the plant G and a transfer function matrix
$\overline{D}$ = a left coprime factor of the plant G and a transfer function matrix
G = the plant transfer function matrix, $G = ND^{-1} = \overline{D}^{-1}\overline{N}$
I = the identity matrix
N = a right coprime factor of the plant G and a transfer function matrix
$\overline{N}$ = a left coprime factor of the plant G and a transfer function matrix
P = a right coprime factor and a transfer function matrix
$\overline{P}$ = a left coprime factor and a transfer function matrix
Q = a right coprime factor and a transfer function matrix
$\overline{Q}$ = a left coprime factor and a transfer function matrix

A. Nassirharand, *Computer-aided Nonlinear Control System Design: Using Describing Function Models*, DOI 10.1007/978-1-4471-2149-7,
© Springer-Verlag London Limited 2012

D.1.1 Multivariable Case

For multivariable systems, consider the following state-space representation:

$$\dot{x}(t) = Ax(t) + Bu(t) \tag{D.3}$$

$$y(t) = Cx(t) + E \tag{D.4}$$

where

A = system matrix, $\in R^{n\times n}$
B = input matrix, $\in R^{n\times p}$
C = output matrix, $\in R^{q\times n}$
E = transmission matrix, $\in R^{q\times p}$
t = time variable
u = vector of inputs of size p
x = vector of state-variables of size n
y = vector of outputs of size q

The coprime factors may be obtained from the following relations (Nett et al. 1984):

$$N = (C - EK)(sI - A_0)^{-1}B + E \tag{D.5}$$

$$D = I - K(sI - A_0)^{-1}B \tag{D.6}$$

$$P = K\left(sI - A_0'\right)^{-1}F \tag{D.7}$$

$$Q = I + K\left(sI - A_0'\right)^{-1}(B - FE) \tag{D.8}$$

$$\overline{N} = C\left(sI - A_0'\right)^{-1}(B - FE) + E \tag{D.9}$$

$$\overline{D} = I - C\left(sI - A_0'\right)^{-1}F \tag{D.10}$$

$$\overline{P} = K(sI - A_0)^{-1}F \tag{D.11}$$

$$\overline{Q} = I + (C - EK)(sI - A_0)^{-1}F \tag{D.12}$$

where

$$A_0 = A - BK \tag{D.13}$$

$$A'_0 = A - FC \tag{D.14}$$

Matrices F and K are determined such that eigenvalues of A_0 and A'_0 have negative real parts; if such F and K exist, then the pairs (A, B) and (A, C) are said to be stabilizable (i.e., all unstable poles are state controllable) and detectable (i.e., all unstable poles are observable), respectively (Skogestad and Postlethwaite 2005).

D.1.2 Single-Variable Case

For single-input single-output systems, it has been shown that the application of the above relations to obtain the coprime factors results in high-order coprime factors (Nassirharand and Patwardhan 1988; Patwardhan 1988; Nassirharand 1993, 2003). Given the plant transfer function, G, a lucid factorization of G is given by (see Vidyasagar 1985):

$$\left\{\begin{array}{l} G = ND^{-1}, G \in R(s), N \in H, D \in H, N = \dfrac{n_0}{m}, \\ n_0 \in \psi, m \in \psi_h, D = \dfrac{d_0}{m}, d_0 \in \psi \end{array}\right\} \tag{D.15*}$$

where

d_0 = the denominator polynomial of G
H = a ring; this ring is the set of all rational functions whose poles lie on the left half of the s-plane
m = Hurwitz polynomial whose poles are in a user-defined algebraic ring H
n_0 = the numerator polynomial of G
$R(s)$ = field of rational functions in s
ψ = the set of all polynomials in s
ψ_h = the set of all Hurwitz polynomials in ψ

Theorem 1 (Nassirharand 2003). Lucid factors of Bezout's identity are acquired by a lucid factorization of the plant, a user-defined algebraic ring, and a set of linear simultaneous algebraic equations whose coefficients form a Sylvester matrix.

Proof. Without loss of generality, bring in a lucid factorization of plant $G(s)$ with degree of $m(s)$, δ_m, equal to max $\{\delta_{n0}, \delta_{d0}\}$; i.e.:

$$G(s) = \frac{N(s)}{D(s)} = \left[\frac{n_0(s)}{m(s)}\right] \cdot \left[\frac{d_0(s)}{m(s)}\right]^{-1} \tag{D.16*}$$

*Source: (Nassirharand 2003) © Elsevier

Now let

$$P(s) = \frac{p_0(s)}{p_1(s)} \tag{D.17*}$$

$$Q(s) = \frac{q_0(s)}{q_1(s)} \tag{D.18*}$$

where $p_0(s) \in \psi$, $p_1(s) \in \psi_h$, $q_0(s) \in \psi$, $q_1(s) \in \psi_h$, $n_0(s) \in \psi$, $d_0(s) \in \psi$, and $m(s) \in \psi_h$. By substituting Eqs. D.15–D.18 into Bezout's identity $\{P(s)N(s) + Q(s)D(s) = 1\}$, we obtain the following equation:

$$p_0(s)n_0(s)q_1(s) + q_0(s)d_0(s)p_1(s) = m(s)q_1(s)p_1(s) \tag{D.19*}$$

let $p_1(s) = q_1(s) = d(s)$ with $d(s) \in \psi_h$. Then, the Bezout's identity becomes:

$$p_0(s)n_0(s) + q_0(s)d_0(s) = m(s)d(s) \tag{D.20*}$$

Note that coprime factors P and Q must at most be proper. For this reason we let $\delta_{p0} = \delta_{q0} = \delta_d$, where the symbol δ is used to denote the degree of a polynomial, and it may be subscripted for clarity. The following relation gives the number of unknown coefficients on the left-hand side of Eq. D.20, denoted T_l:

$$T_l = 2\delta_d + 2 \tag{D.21*}$$

And the number of unknown coefficients on the right-hand side of Eq. D.20, denoted T_r, is given by:

$$T_r = \delta_m + \delta_d + 1 \tag{D.22*}$$

In order for equality (D.20) to hold, we must have:

$$T_l = T_r \Rightarrow \delta_d = \delta_m - 1 \equiv l \tag{D.23*}$$

Let

$$n_0(s) = n_0^0 + n_0^1 s + n_0^2 s^2 + \cdots \tag{D.24*}$$

$$d_0(s) = d_0^0 + d_0^1 s + d_0^2 s^2 + \cdots \tag{D.25*}$$

$$p_0(s) = p_0^0 + p_0^1 s + p_0^2 s^2 + \cdots \tag{D.26*}$$

---

*Source: (Nassirharand 2003) © Elsevier

$$q_0(s) = q_0^0 + q_0^1 s + q_0^2 s^2 + \cdots \tag{D.27*}$$

$$m(s) = m^0 + m^1 s + m^2 s^2 + \cdots \tag{D.28*}$$

$$d(s) = d^0 + d^1 s + d^2 s^2 + \cdots \tag{D.29*}$$

Select $m(s) \in \psi_h$ and $d(s) \in \psi_h$ such that coprime pairs $[P, Q] \in H$ and $[N, D] \in H$; then, substitute the above polynomials of known degrees into Bezout's identity (or Eq. D.20); expand identity terms and equate terms of equal power in complex frequency. The following equations would be obtained:

$$n_0^0 p_0^0 + d_0^0 q_0^0 = m^0 d^0 \tag{D.30*}$$

$$n_0^1 p_0^0 + n_0^0 p_0^1 + d_0^1 q_0^0 + d_0^0 q_0^1 = m^0 d^1 + m^1 d^0 \tag{D.31*}$$

$$\begin{aligned} & n_0^l p_0^0 + n_0^{l-1} p_0^1 + \cdots n_0^0 p_0^l + d_0^l q_0^0 + d_0^{l-1} q_0^1 + \cdots + d_0^0 q_0^l \\ & = m^0 d^l + m^1 d^{l-1} + \cdots + m^l d^0 \end{aligned} \tag{D.32*}$$

$$\begin{aligned} & n_0^{l+1} p_0^0 + n_0^l p_0^1 + \cdots + n_0^1 p_0^l + d_0^{l+1} q_0^0 + d_0^l q_0^1 + \cdots + d_0^1 q_0^l \\ & = m^0 d^{l+1} + m^1 d^l + \cdots + m^{l+1} d^0 \end{aligned} \tag{D.33*}$$

$$\begin{aligned} & n_0^{2l+1} p_0^0 + n_0^{2l} p_0^1 + \cdots + n_0^{l+1} p_0^l + d_0^{2l+1} q_0^0 + d_0^{2l} q_0^1 + \cdots \in + d_0^{l+1} q_0^l \\ & = m^0 d^{2l+1} + m^1 d^{2l} + \cdots + m^{2l+1} d^0 \end{aligned} \tag{D.34*}$$

By noting that $\{n_0^k = 0, k \in i := (l+2, l+3, ..., 2l+1)\}$, $\{d_0^k = 0, k \in i : = (l+2, l+3, ..., 2l+1)\}$, $\{m^k = 0, k \in i := (l+2, l+3, ..., 2l+1)\}$, and $\{d^k = 0, k \in i := (l+1, l+2, ..., 2l+1)\}$, the above set of equations in matrix form follows.

$$SX = B' \tag{D.35*}$$

where

$$S = \begin{bmatrix} S_{11} & S_{12} \\ S_{21} & S_{22} \end{bmatrix} \tag{D.36*}$$

*Source: (Nassirharand 2003) © Elsevier

$$S_{11} = \begin{bmatrix} n_0^0 & 0 & \cdots & 0 & 0 \\ n_0^1 & n_0^0 & \ddots & \ddots & \vdots \\ \vdots & n_0^1 & \ddots & 0 & 0 \\ \vdots & \vdots & \ddots & n_0^0 & 0 \\ n_0^l & n_0^{l-1} & \cdots & n_0^1 & n_0^0 \end{bmatrix} \quad S_{12} = \begin{bmatrix} d_0^0 & 0 & \cdots\cdots & 0 \\ d_0^1 & d_0^0 & \ddots & 0 & 0 \\ \vdots & \ddots & \ddots & 0 & 0 \\ \vdots & \ddots & \ddots & d_0^0 & 0 \\ d_0^l & d_0^{l-1} & \cdots & d_0^1 & d_0^0 \end{bmatrix}$$

$$S_{21} = \begin{bmatrix} n_0^{l+1} & n_0^l & \cdots & \cdots & n_0^1 \\ 0 & n_0^{l+1} & n_0^l & \ddots & n_0^2 \\ \vdots & 0 & \ddots & \ddots & \vdots \\ \vdots & \ddots & \ddots & n_0^{l+1} & n_0^l \\ 0 & 0 & \cdots & 0 & n_0^{l+1} \end{bmatrix} \quad S_{22} = \begin{bmatrix} d_0^{l+1} & d_0^l & \cdots & \cdots & d_0^1 \\ 0 & d_0^{l+1} & \ddots & \ddots & d_0^2 \\ \vdots & 0 & \ddots & d_0^l & \vdots \\ \vdots & \ddots & \ddots & d_0^{l+1} & d_0^l \\ 0 & 0 & \cdots & 0 & d_0^{l+1} \end{bmatrix}$$

$$B' = \begin{bmatrix} m^0 d^0 \\ m^0 d^1 + m^1 d^0 \\ \vdots \\ \vdots \\ m^0 d^l + m^1 d^{l-1} + \ldots + m^l d^0 \\ m^1 d^l + m^2 d^{l-1} + \ldots + m^{l+1} d^0 \\ \vdots \\ \vdots \\ \vdots \\ m^{l+1} d^l \end{bmatrix} \tag{D.37*}$$

In the above equation, S is known, B' will be chosen by the user, and X is unknown and will be determined from Eq. D.35. Notice that we started with a lucid factorization of the plant that was followed by selection of the coprime pairs $[N, D]$ and $[P, Q]$ in a user-defined algebraic ring, and finally, it was shown that remaining unknowns may be solved via a linear matrix equation of the form given by Eq. D.35 whose coefficient matrix, S, is a Sylvester matrix. Notice that Sylvester matrix is always nonsingular provided that $n_0(s) \in \psi$ and $d_0(s) \in \psi$ do not have any common factors.

*Source: (Nassirharand 2003) © Elsevier

Remark 1. In order to obtain the lower-order coprime factors that are solutions to the Bezout's identity, we must select $m(s)$ to enforce either $N(s) \in H$ or $D(s) \in H$ to have left-half plane pole/zero cancellations.

D.1.3 Software

A MATLAB® function for computation of coprime factors given by Eqs. D.5–D.12 is developed in Teh (2010). With reference to the work presented in Nassirharand and Patwardhan (1988), a much simpler software in the MATLAB® environment for single-variable systems is developed in Nassirharand (2003). The m-file listings are given in Fig. D.1.

The corresponding MATLAB® function is of the following form:

$$[N, D, P, Q] = Lucid\_factor(G, mpoles, dpoles)$$

where N, D, P, Q, and G are defined as described in Sect. 3; $mpoles$ is an array of size $\delta_m = \max(\delta_{n0}, \delta_{d0})$ containing the desired poles of plant lucid factors, and $dpoles$ is an array of size $\delta_d = \delta_m - 1$ containing the desired poles of the coprime pair $[P, Q] \in H$.

The software first obtains the lucid factorization of the plant followed by determination of the Sylvester coefficient matrix, S; then, the input matrix B' is formed, and finally the unknown vector X is solved for. The last step of the software is to enforce any left-half plane pole-zero cancellations that may exist.

D.2 DRLCD Using Simultaneous Stabilization Theory

D.2.1 Theory and Algorithm

The original idea for design of dual-range controllers for nonlinear systems (Taylor 1983) calls for the application of simultaneous stabilization theory of Vidyasagar and Viswanadham (1982). In this section, all controllers that simultaneously stabilize the closed-loop systems corresponding to the two linear system models of the nonlinear plant are parameterized. This involves developing a mathematical expression for all stabilizing controllers in terms of the coprime factors of linear system models and another as yet unknown stable nth-order transfer function denoted $r(s)$.

The term *parameterization* arises from the fact that the class of all stabilizing controllers is expressed in terms of the *parameter* $r(s)$. Such a parameterization provides a logical means for obtaining the desired minimum-sensitivity controller. The following is single-input/single-output version of the simultaneous stabilization theory.

```
%
% FUNCTION Lucid_factor SOLVES FOR COPRIME FACTORS
% N,D,P,Q GIVEN BY SINGLEVARIABLE VERSION OF EQUATION
%(D.1) THE INPUTS ARE THE PLANT TRANSFER FUNCTION, G,
% DESIRED POLES OF IDENTICAL DENOMINATOR  POLYNOMIALS OF
% N AND D IN EQUATION (D.16) (i.e., mpoles) AND THE
% IDENTICAL POLES OF P AND Q (see Equations (D.17) AND
% (D.18)) LENGTH OF mpoles MUST BE ONE LESS THAN LENGTH
% OF dpoles.  THE OUTPUTS ARE THE COPRIME FACTORS N, D,
% P, AND Q
%
function [N,D,P,Q]=Lucid_factor(g,mpoles,dpoles)
%
% OBTAIN A LUCID FACTORIZATION OF PLANT g
[N,D]=lucid(g,mpoles);
%
[n0,m]=tfdata(N,'v');[d0,m]=tfdata(D,'v');
l=length(m)-2;
%
% OBTAIN THE S MATRIX DEFINED BY EQUATION (D.36), n4=S
n4=getn4(n0,d0,l);
%
d=poly(dpoles);
% OBTAIN THE B-prime VECTOR IN EQUATION (D.37).
% b=B-prime.
b=getb(m,d);
%
% COMPUTE COPRIME FACTORS P AND Q
%
[P,Q]=getpq(n4,b,l,d);
tol=0.001;N=minreal(N,tol);D=minreal(D,tol);
P=minreal(P,tol);Q=minreal(Q,tol);
%
%************** END OF FUNCTION Lucid_factor **********
```

Fig. D.1 Listing of the m-files for lucid factorization and solution to the Bezout's identity. Source: (Nassirharand 2003) © Elsevier

```
% FUNCTION getb COMPUTES MATRIX B-prime GIVEN BY EQUATION
% (D.35)
% THE INPUTS ARE POLYNOMIALS m AND d
% THE OUTPUT IS VECTOR B-prime.  b=B-prime.
function b=getb(m,d);
% COMPUTE m.d
md=conv(m,d);
len=length(md);
j=0;
for i=len:-1:1
   j=j+1;
   b(j)=md(i);
end
b=b(:);
%************* END OF FUNCTION getb ****************
% FUNCTION getn4 COMPUTES MATRIX S IN EQUATION (D.35)
% THE INPUTS ARE POLYNOMIALS n0 and d0 GIVEN BY EQUATIONS
% (D.24) AND (D.25).  ALSO SEE EQUATION (D.16).
function n4=getn4(n0,d0,l)
% COMPUTE S11, S12, S21, AND S22.  n11=S11, n12=S12,
% n21=S21, AND n22=S22.  SEE EQUATION (D.35).
n11=stanley(n0,l);n12=stanley(d0,l);
n21=toeplitz(n0,l);n22=toeplitz(d0,l);
n4=[n11 n12;n21 n22];
%****************** END OF FUNCTION getn4 ***************
% FUNCTION getpq COMPUTES COPRIME FACTORS P AND Q
% THE INPUTS ARE n4=S (see Equation (D.36)), b=B-prime
% (see Equation (D.37), AND POLYNOMIAL d (see Equation
% (D.29))
function [P,Q]=getpq(n4,b,l,d);
% COMPUTE X IN EQUATION (D.35)
x=n4\b;
j=0;
for i=l+1:-1:1;j=j+1;p(j)=x(i);end
j=0;
for i=2*l+2:-1:l+2;j=j+1;q(j)=x(i);end
P=tf(p,d);Q=tf(q,d);
%****************** END OF FUNCTION getpq **************
```

Fig. D.1 (continued)

```
%
% FUNCTION lucid COMPUTES COPRIME FACTORS N AND D USING
% EQUATION (D.16).  ALSO SEE VIDYASAGAR 1985.
% THE INPUTS ARE PLANT TRANSFER FUNCTION G AND THE
% DESIRED POLES OF N AND D.  SEE REMARK IN SECTION D.1.2
% THE OUTPUTS ARE COPRIME FACTORS N AND D
function [n,d]=lucid(g,poles)
[n0,d0]=tfdata(g);
m=poly(poles);n=tf(n0,m);d=tf(d0,m);
%************* END OF FUNCTION lucid ******************
% FUNCTION stanley IS USED TO COMPUTE S11 AND S12 GIVEN
% BY EQUATION (D.36)
function s =stanley(n0,l)
for i=1:l+1;for j=1:l+1;s(i,j)=0;end;end
for i=1:l+1
   for j=1:l+1
      if(i==j)
         s(i,j)=n0(l+2);
      elseif(i>j)
         s(i,j)=n0(l+2-i+j);
      end
   end
end
%***************     END     OF     FUNCTION     stanley
***************
%
% FUNCTION toeplitz IS USED TO COMPUTE S21 AND S22 GIVEN
% BY EQUATION (D.36)
function s =toeplitz(n0,l)
for i=1:l+1;for j=1:l+1;s(i,j)=0;end;end
for i=1:l+1
   for j=1:l+1
      if(i==j)
         s(i,j)=n0(1);
      elseif(i<j)
         s(i,j)=n0(1+j-i);
      end
   end
end
%*************** END OF FUNCTION toeplitz *************
```

Fig. D.1 (continued)

Theorem 2. The class of all controllers that stabilize a plant $G(s)$ may be parameterized in terms of a stable linear transfer function, $r(s)$, as follows:

$$C(s) = \frac{P(s) + r(s)D(s)}{Q(s) - r(s)N(s)} \tag{D.38}$$

where

$C(s)$ = the stabilizing controller
$r(s)$ = any stable transfer function parameter to be selected by the designer; may be selected to achieve certain performance criteria,

and the rest of the symbols are defined as in Eq. D.1. See Youla et al. (1976) for proof.

The parameterization of all controllers that stabilize two stable linear system models is given next. Let the pair $[N_i(s), D_i(s)]$ correspond to the stable coprime factors of $G_i(s)$, $i = 0, 1$ and designate the corresponding factors that satisfy the Bezout's identify by the stable coprime pair $[P_i(s), Q_i(s)]$.

Theorem 3. The set of all controllers that simultaneously stabilize $G_0(s)$ and $G_1(s)$ is given by:

$$C_{ss}(s) = \frac{P_0(s) + \overline{r}(s)D_0(s)}{Q_0(s) - \overline{r}(s)N_0(s)}, \tag{D.39}$$

where $\overline{r}(s)$ belongs to the set of all *stable* controllers that stabilize $G_a(s)$

$$G_a(s) = \frac{-N_0(s)D_1(s) + D_0(s)N_1(s)}{Q_0(s)D_1(s) + P_0(s)N_1(s)} \tag{D.40}$$

The set of all controllers that stabilize $G_a(s)$ is given by Theorem 2; the condition that $\overline{r}(s)$ must itself be stable (it is said that the plant G_a must be *strongly* stabilizable) must be imposed in addition. See Vidyasagar and Viswanadham (1982) for proofs.

The following algorithm (Patwardhan 1988; Nassirharand 1991; Ujile and Nassirharand 2010), which is based on the simultaneous stabilization theory presented above, may be applied to arrive at a dual-range controller.

Algorithm D.2

1. Define the desired performance measures for the final closed-loop system. For example, one could specify the desired percentage overshoot, rise time, settling time, undershoot, gain margin, and/or phase margin.
2. Identify a reference linear model whose characteristics are similar to those of the performance measured defined in the previous step.

3. Set the denominator of the design parameter equal to that of the reference linear model identified in Step 2. This choice for the design parameter denominator is based on dominant pole placement ideas.
4. Use the *Lucid_factor* MATLAB® function (see Fig.D.1) and compute the coprime factors of the plant transfer functions as well as the coprime factors that are solutions to the Bezout's identity.
5. Compute the auxiliary plant, $G_a(s)$, given by Eq. D.40. Then, classify the set of all controllers that stabilize the auxiliary plant in terms of a function parameter, $R_a(s)$, as below (see Theorem 2):

$$C_a(s) = \frac{P_a(s) + R_a(s)D_a(s)}{Q_a(s) - R_a(s)N_a(s)} \tag{D.41}$$

where

D_a = a coprime factor of the plant G_a and a transfer function
N = a coprime factor of the plant G_a and a transfer function
P_a = a right coprime factor that is a solution to the Bezout's identity and a transfer function
Q_a = a right coprime factor that is a solution to the Bezout's identity and a transfer function

The simultaneously stabilizing controller, $C_{ss}(s)$, is given by Eq. D.39. InEq. D.39, $\overline{r}$ is given by $C_a(s)$ for which $G_a(s)$ is strongly stabilizable. Substituting $C_a(s)$ from Eq. D.41 into Eq. D.39 as $\overline{r}$ solves the simultaneous stabilization problem.

6. Search for the optimum coefficients of the numerator polynomial of $R_a(s)$ using dominant pole placement ideas coupled with optimization to yield the desired simultaneously stabilizing controller that forces the closed-loop system to behave in a desired way. To achieve this, the following objective function is minimized using a simplex optimization routine as implemented by *fminsearch* function of MATLAB (see Nelder and Mead 1965 for a Simplex Algorithm):

$$F = \beta F_0 + (1 - \beta) F_1 \tag{D.42}$$

where

$$F_j(x) = \alpha\left[\left\{y^j_{\max} - y^d_{\max}\right\}^2 + \langle y^j_{\min}\rangle^2\right] + \sum_i \left(y^j_i - y^d_i\right)^2, j = 0, 1 \tag{D.43}$$

$$\langle y^j_{\min}\rangle^2 = \begin{cases} \left(y^j_{\min}\right)^2 \text{ if } y^j_{\min} < 0 \\ 0 \text{ if } y^j_{\min} > 0 \end{cases} \tag{D.44}$$

d = a superscript denoting the desired value(s)
y = the unit step response; may be subscripted and/or superscripted for clarity
α = a weighting factor
β = a weighting factor

The terms in brackets of Eq. D.43 represent the errors in the overshoot and undershoot specifications, respectively. The remaining term in Eq. D.43 represents the squared error between the actual and desired closed-loop response. The objective function has two weighting factors, α and β. Weighting factor α is a trade-off between the overshoot and settling time conditions, while β is a trade-off between the closed-loop responses of the two different plants.

The vector of independent variables is comprised of the coefficients of the numerator polynomial of $R_a(s)$. Initial values for these coefficients must be assumed.

9. Validate design using a digital simulation.

D.2.2 *Software*

There exist two pieces of software based on Algorithm D.2. One is in the FORmula TRANslation language whose listing is given in Patwardhan (1988). The other is in terms of MATLAB® functions developed in Ujile (2010). The flow chart of the MATLAB® version is given in Figs. D.2 and D.3 (Ujile and Nassirharand 2010).

D.2.3 *Example Problems*

D.2.3.1 A Linear System with Integrity

Consider a linear system with integrity whose dynamic behavior at two different conditions is given by $G_0(s)$ and $G_1(s)$ as follows:

$$G_0(s) = \frac{1}{4\,(s+2)} \tag{D.45}$$

$$G_1(s) = \frac{3.0}{4\,(s-2)} \tag{D.46}$$

Fig. D.2 Flowchart for the main function of DRLCD. Source: (Ujile and Nassirharand 2010) © Hyperssciences

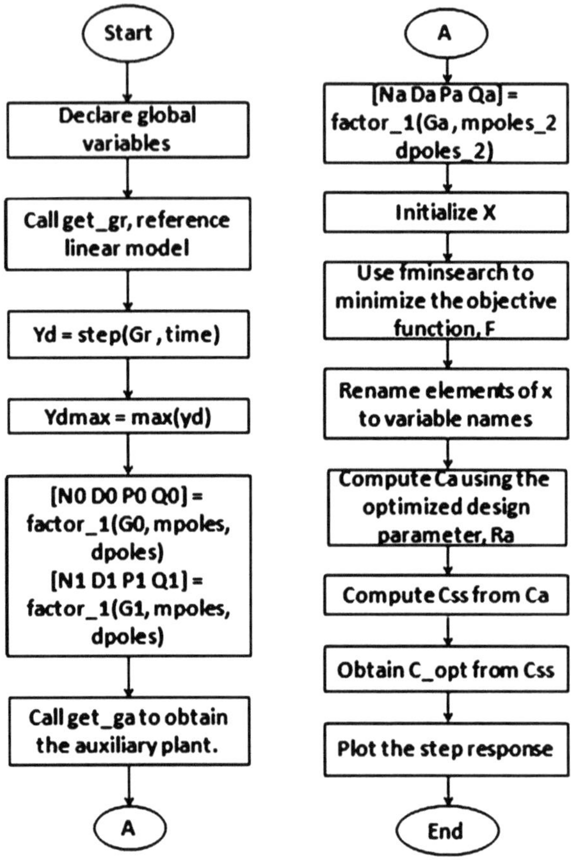

This example problem was originally studied in Nassirharand (1986), and later in Patwardhan (1988), Nassirharand (1991), Ujile and Nassirharand (2010), Teh (2010), and Nassirharand (2010b). It is desired to design a dual-range controller so the closed-loop system comprised of the designed controller and the two linear models above would mimic a second-order linear system with a percent overshoot of 37 and a settling time of 0.3 s, i.e.:

$$h_{y,u}^{D} = \frac{1867.9}{s^2 + 26.5s + 1867.9} \tag{D.47}$$

The plant $G_0(s)$ is stable. Hence, we choose $N_0(s) = G_0(s)$ and $D_0(s) = 1$. The auxiliary plant $G_a(s)$ is given by the following equation:

$$G_a(s) = \frac{s+4}{2\,(s^2-4)} \tag{D.48}$$

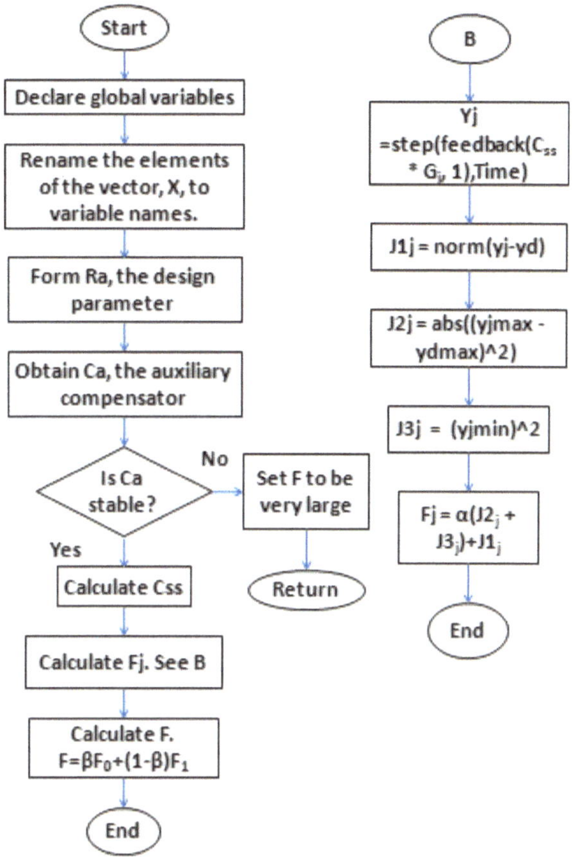

Fig. D.3 Flowchart for computation of the objective function for DRLCD. Source: (Ujile and Nassirharand 2010) © Hypersciences

Since $G_a(s)$ has the parity interlacing property (PIP), a compensator that strongly stabilizes $G_a(s)$ exists (a plant has PIP if the number of poles of the plant between every pair of real zeros on the right half plane is even). To find the auxiliary compensator $C_a(s)$, first the coprime factors of $G_a(s)$ are formed as given below:

$$N_a(s) = \frac{s+4}{s^2+4s+4} \tag{D.49}$$

$$D_a(s) = \frac{2\,(s-2)}{(s+2)} \tag{D.50}$$

For the above coprime factors, the solution to the Bezout's identity is obtained as follows:

$$P_a(s) = 2.6667 \tag{D.51}$$

$$Q_a(s) = \frac{s+3.33}{2\,(S+2)} \tag{D.52}$$

Hence, the strongly stabilizing controller is given by:

$$C_a(s) = \frac{r_{ad}\left(21.3s^3 + 127.9s^2 + 255.9s + 170.6\right) + r_{an}\left(16.0s^3 + 32.0s^2 - 64.0s - 128.0\right)}{r_{ad}\left(4.0s^3 + 23.9s^2 + 47.9s + 31.9\right) - r_{an}\left(12.0s^2 + 48.0s + 48.0\right)} \tag{D.53}$$

where

r_{ad} = the denominator polynomial of the transfer function parameter $R_a(s)$ in Eq. D.41
r_{an} = the numerator polynomial of the transfer function parameter $R_a(s)$ in Eq. D.41

To maximize the number of the independent design variables for optimization, we set the degree of the numerator polynomial of the parameter $R_a(s)$ equal to that of its denominator. Hence,

$$r_{an}(s) = a's^2 + b's + c' \tag{D.54}$$

The design variable c' in Eq. D.54 is selected to have a zero steady-state error due to a step input, i.e., $c' = 1243.36$. Optimization is used to select the design variables a′ and b' in the following fashion. The value of $\beta = 0.5$ is chosen to give the same weight to the responses due to both plants. Keeping β fixed at 0.5 with initial values $a' = b' = 1.0$, and $\alpha = 1000$, the two software mentioned in Sect. D.2.2 produce the following compensators:

$$C_{ss}^1(s) = \frac{39.4s^3 + 2518s^2 + 1.13e4s + 1.055e4}{s^3 + 3.297s^2 + 102.6s} \tag{D.55}$$

$$C_{ss}^2(s) = \frac{15.06s^2 + 2723s + 3834}{s^2 + 19.3s} \tag{D.56*}$$

where

C_{ss}^1 = the controller obtained using the FORTRAN software after simplification and model reduction
C_{ss}^2 = the controller obtained using the MATLAB® software after simplification and model reduction

Unlike the FORTRAN solution, in MATLAB® solution the value c' is also determined by optimization; its initial value is set to 1; also, poles of the coprime factors are set to −2. For comparison purposes, the software based on direct optimization approach is also applied; the following controller is obtained:

*Source: (Ujile and Nassirharand 2010) © Hypersciences

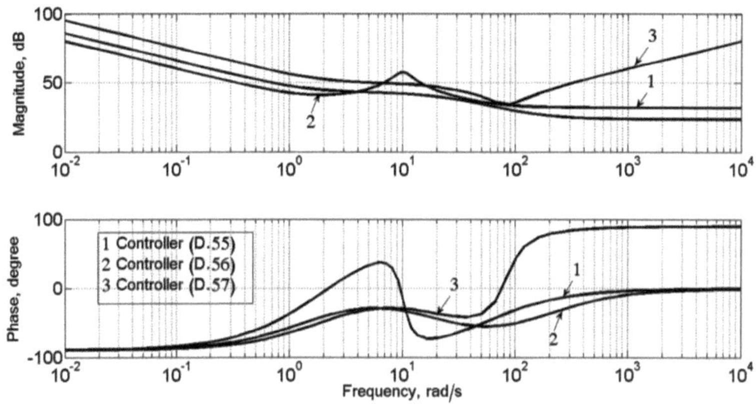

Fig. D.4 Comparison of frequency response of the three simultaneous stabilizing controllers

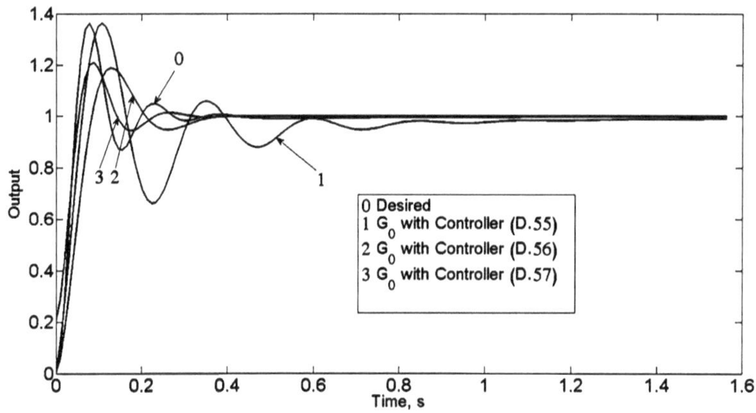

Fig. D.5 Step-response comparison of the performance of different simultaneously stabilizing controllers with plant G_0 (linear system with integrity)

$$C_{ss}^{3}(s) = \frac{1.024s^3 + 54.98s^2 + 7395s + 1.416e4}{s^2 + 24.26s} \tag{D.57}$$

where

C_{ss}^{3} = the controller obtained using the MATLAB® software using direct optimization (Teh 2010) after simplification

The frequency responses of all three controllers may be compared by examining Fig. D.4. The unit step responses of the closed-loop systems comprised of both plants (see Eqs. D.45 and D.46) with all three controllers along with that of the desired are depicted in Figs. D.5 and D.6.

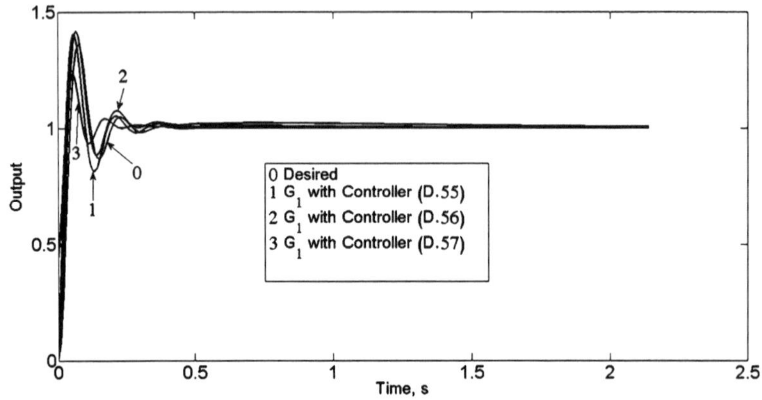

Fig. D.6 Step-response comparison of the performance of different simultaneously stabilizing controllers with plant G_1 (linear system with integrity)

D.2.3.2 A Nonlinear Servo System

The application of the simultaneous stabilization theory to a nonlinear plant is investigated in this section. Consider the nonlinear plant that is also studied in Sects. 2.2.1.1, 4.4, 5.4.1, 5.4.2, and 5.4.3. The describing function models of this nonlinear plant are depicted in Figs. 2.5, 2.6, 2.7, and 2.8. The two describing function models whose gain characteristics bound those of all the others in class are identified to be as follows (Nassirharand et al. 1988; Nassirharand 1988):

$$G_0(s) = \frac{1.231 + 0.003989s}{s\,(0.05475 + 0.003761s)} \tag{D.58}$$

$$G_1(s) = \frac{1.000 + 0.04031s}{s\,(0.22521 + 0.23641s)} \tag{D.59}$$

The desired reference linear model is given as follows:

$$h^D_{y,u} = \frac{1953}{s^2 + 26.67s + 1953} \tag{D.60}$$

The MATLAB® implementation of Algorithm D.2 with the same starting solution as the previous example and the same location of the coprime factor poles with $\beta = 0.1$ is executed to obtain the following controller:

$$C^4_{ss} = \frac{228s^4 + 3441s^3 + 1.051e4s^2 + 1.424e4s + 7386}{s^4 - 5.166s^3 + 643.7s^2 - 668.1s + 1484} \tag{D.61}$$

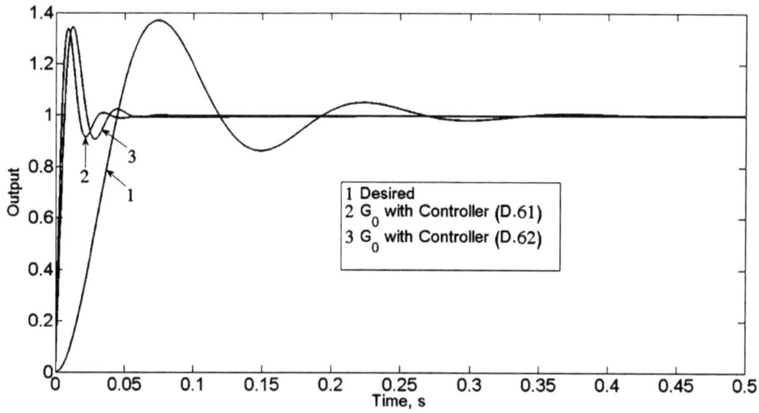

Fig. D.7 Step-response comparison of the performance of different simultaneously stabilizing controllers with plant G_0 (nonlinear system)

The particular reason for selecting $\beta = 0.1$ is because the nonlinear effects (i.e., Coulomb friction) are more pronounced at small signals, and hence, it is decided to pay more attention to the model given by $G_1(s)$. This value for β is also selected in Nassirharand (1991).

The controller obtained by direct optimization (Teh 2010) is given as below (see Chap. 4):

$$C_{ss}^5(s) = \frac{3.1671s + 1.1469}{0.0229s + 0.1089} \tag{D.62}$$

The unit step-response plots of both controllers with both linear approximations are depicted in Figs. D.7 and D.8.

The unit step-response plots of $C_{ss}^5(s)$ with the nonlinear plant are depicted in Fig. D.9. In this example, although $C_{ss}^4(s)$ stabilizes both $G_0(s)$ and $G_1(s)$, it does not stabilize the actual nonlinear system (see Fig. D.10). Hence, the original call to use the simultaneous stabilization theory of linear systems for nonlinear control system design becomes questionable.

D.2.4 Discussion

The primary objectives of this appendix were to demonstrate the following two major problems with the application of the simultaneous stabilization theory of linear systems to nonlinear systems:

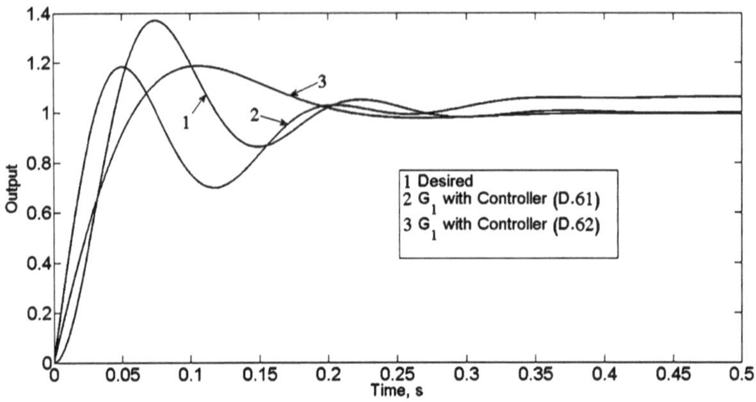

Fig. D.8 Step-response comparison of the performance of different controllers with plant G_1 (nonlinear system)

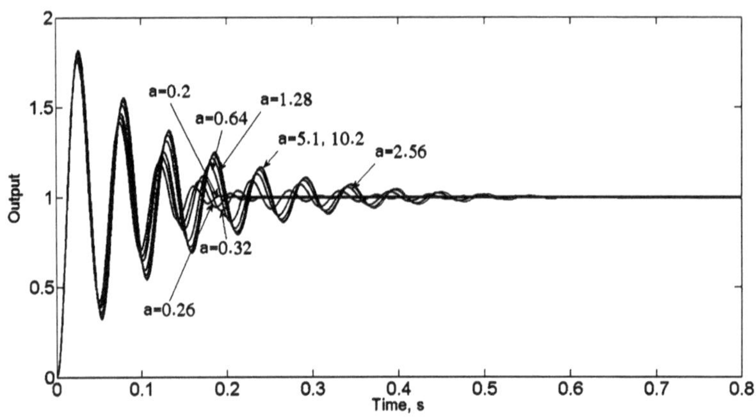

Fig. D.9 Normalized step-response plots of the closed-loop system with the nonlinear plant and the simultaneously stabilizing controller (D.62)

1. The simultaneously stabilizing controllers tend to be of a high order.
2. The simultaneously stabilizing dual-range controllers may not stabilize the actual nonlinear system.

These goals have been met. In Sect. D.2.3.1, a linear system with integrity is considered. In Sect. D.2.3, a nonlinear example problem is considered, and for this problem it is demonstrated that simultaneously stabilizing dual-range controller obtained from the application of the simultaneous stabilization theory of linear systems is of a much higher order than that obtained using a direct optimization approach. It is also demonstrated that, unlike the latter dual-range controller, the former dual-range controller does not stabilize the actual nonlinear plant.

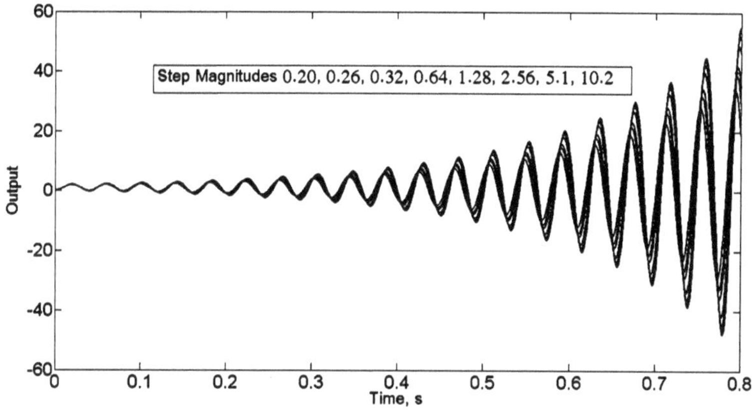

Fig. D.10 Normalized step-response plots of the closed-loop system with the nonlinear plant and the simultaneously stabilizing controller (D.61) – unstable!

It could certainly be argued that the direct optimization approach (Teh 2010; Nassirharand et al. 1988; Nassirharand 1991; Patwardhan 1988) described in Chap. 4 could also result in controllers that would stabilize the two linear models but not the actual nonlinear plant. In such cases, one should consider design of a nonlinear controller as described in Chap. 5. However, application of the direct optimization approach to various nonlinear problems in aerospace and servomechanism has not shown such a characteristic (see Teh 2010; Teh and Nassirharand 2010b, c; Chap. 4; and Sect. D.2.3.1).

References

Abbassi N (2003) Controller design for simultaneous control of combustion chamber pressure and mixture ratio of a specific liquid propellant engine. M.S. Thesis Department of Electrical and Computer Engineering, Amir Kabir University, Tehran

Al-Kindi SSA, Nassirharand A (2010) State-variable control of quadrotor helicopter. In: Proceedings of international conference on advances in mechanical engineering, Shah Alam

Atherton DP (1970) Comment on describing function inversion. Electron Lett 6:779–780

Atherton DP (1975) Nonlinear control engineering. van Nostrand Reinhold, London

Atherton DP, Benouarets M, Nanka-Bruce O (1993) Design of nonlinear PID controllers for nonlinear plants. Proc IFAC World Congr Sydney 3:355–358

Bengea SC, Li X, Decarlo RA (2004) Combined controller-observer design for uncertain time delay systems with application to engine idle speed control. ASME Trans J Dyn Syst Meas Control 126:772–780

Besancon-Voda A, Blaha P (2002) Describing function approximation of a two-relay system configuration to Coulomb friction identification. Control Eng Pract 10:655–668

Bongard JC, Lipson H (2005) Nonlinear system identification using coevolution of models and tests. IEEE Trans Evol Comput 9:361–384

Chen CT (1984) Linear system theory and design. Holt, Rinehart, and Winston, New York

Chen CF, Shieh LS (1970) An algebraic method for control systems design. Int J Control 1:717–739

Colgren RD (2004) Applications of robust control to nonlinear systems. American Institute of Aeronautics and Astronautics, Reston

D'Azzo JJ, Houpis CH (1966) Feedback control system analysis and synthesis, 2nd edn. McGraw-Hill Book Company, New York

Dhupia JS, Ulsoy AG, Katz R et al (2008) Experimental identification of the nonlinear parameters of an industrial translational guide for machine performance evaluation. J Vib Control 14:645–668

Doyle JC, Glover K, Khargonekar PP, Francis BA (1989) State-space solutions to standard H_2 and H_∞ control problems. IEEE Trans Auto Control 34:831–847

Garvin C, Mathew A (1996) The application of the method of simultaneous stabilization to the control of a nonlinear servo valve. IEEE Trans Control Syst Technol 4:654–664

Ge HW, Qian F et al (2008) Identification and control of nonlinear systems by a dissimilation particle swarm optimization-based Elman neural-network. Nonlin Anal Real World Appl 9:1345–1360

Gelb A, Vander Velde WE (1968) Multiple-input describing functions and nonlinear system design. McGraw-Hill, New York

A. Nassirharand, *Computer-aided Nonlinear Control System Design: Using Describing Function Models*, DOI 10.1007/978-1-4471-2149-7,

© Springer-Verlag London Limited 2012

Govindan P A/L R (2011) Controller design for a 2 degree-of-freedom robot arm. BEng Thesis, Faculty of Engineering, The University of Nottingham, Semenyih, Selangor
Haber R, Keviczky L (1999) Nonlinear system identification-input–output modeling approach. Kluwer Acaemic Publishers, Dordrecht
Horowitz IM (1976) Synthesis of feedback systems with nonlinear time-varying uncertain plant to satisfy quantitative performance specifications. Proc IEEE 64:123–130
Hunt LR, Su R, Meyer G (1987) Global transformation of nonlinear systems. IEEE Trans Auto Control 28:24–30
John J (2001) Nonlinear continuous system identification by means of multiple integration II. Acta Polytechnica 41:64–67
Karimi H, Nassirharand A (2006) Application of a simulation algorithm for dynamic and nonlinear analysis of a liquid propellant engine. J Aerosp Sci Technol 3:23–30
Karimi H, Nassirharand A, Beheshti M (2003) Dynamic and nonlinear simulation of liquid propellant engines. AIAA J Propuls Power 19:938–944
Kerschen G, Worden K, Vakakis AF et al (2006) Past, present and future of nonlinear system identification in structural dynamics. Mech Syst Signal Process 20:505–592
Kim MS, Chung SC (2006) Friction identification of ball-screw driven servomechanisms through the limit cycle analysis. Mechatronics 16:131–140
Kuo BC (1982) Automatic control systems, 4th edn. Prentice-Hall, Inc., Englewood Cliffs
Li SC, Nassirharand A (2010) A single-range controller design procedure for one degree-of-freedom servo systems with experimental results. In: Proceedings of international conference on advances in mechanical engineering, Shah Alam
Li SC, Nassirharand A (2011) Non-linear proportional-integral-derivative synthesis for unstable non-linear systems using describing function inversion with experimental verification. Proc Inst Mech Eng Part I J Systems and Control Eng (in press)
Li L, Yang Y, Peng H (2010) Fuzzy system identification via chaotic ant swarm. Chaos Solitons Fractiles 41:401–409
Lyashevskiy S, Chen Y (1997) Identification of nonlinear system parameters using describing function approach. Proc Am Control Conf 3:1944–1945
Mousavi Firdeh SR, Nassirharand A, Abbassi N et al (2006) A systematic single-range controller synthesis procedure for nonlinear and multivariable liquid propellant engines. Aerosp Sci Technol 10:392–401
Nanka-Bruce O, Atherton DP (1990) Design of nonlinear controllers for nonlinear plants. Proc IFAC Congr Tallinn 6:75–80
Nassirharand A (1986) Controller design for nonlinear systems based on simultaneous stabilization theory and describing function models. PhD Thesis, School of Mechanical and Aerospace Engineering, Oklahoma State University, Stillwater, Oklahoma
Nassirharand A (1987) Input/output characterization of highly nonlinear systems. Adv Eng Softw 9:129–133
Nassirharand A (1988) Identification of frequency domain models for nonlinear systems. Adv Eng Softw 10:195–201
Nassirharand A (1991) Design of dual-range linear controllers for nonlinear systems. ASME Trans J Dyn Syst Meas Control 113:590–596
Nassirharand A (1993) Factorization approach to control system synthesis. AIAA J Guid Control Dyn 16:402–405
Nassirharand A (2003) Computation of lucid factors for Bezout identity. Adv Eng Softw 34:527–531
Nassirharand A (2008a) A new software tool for design of linear compensators. Adv Eng Softw 39:132–136
Nassirharand A (2008b) Controller synthesis procedure for linear unstable plants. Int J Automat Control 2:137–149
Nassirharand A (2008c) Control system design of multivariable non-linear space robots. Aircr Eng Aerosp Technol 80:27–34

Nassirharand A (2009a) MATLAB software for inversion of describing functions. Adv Eng Softw 40:600–606
Nassirharand A (2009b) Computer-aided design of classical linear compensators with application to aerospace. J Syst Eng Electron 20:1373–1377
Nassirharand A (2010a) Nonlinear controller synthesis for SIMO systems with application to cruise missile. ASCE J Aerosp Eng 23:105–110
Nassirharand A (2010b) Controller design for nonlinear systems. LAP Lambert Academic Publishing, Saarbrucken
Nassirharand A (2011) Computer-aided design of P or PI plus rate feedback compensators for linear systems. Adv Eng Software 42:1035–1040
Nassirharand A, Housseini SA (2009) Tracking and decoupling of multivariable nonlinear systems. Int J Model Identif Control 6:341–348
Nassirharand A, Karimi H (2002) Input/output characterization of highly nonlinear multivariable systems. Adv Eng Softw 33:825–830
Nassirharand A, Karimi H (2004a) Design of a single-range controller for the pressure control of a combustion chamber. Scientia Iranica 11:153–158
Nassirharand A, Karimi H (2004b) Controller synthesis methodology for multivariable nonlinear systems with application to aerospace. ASME Trans J Dyn Syst Meas Control 126:598–607
Nassirharand A, Karimi H (2004c) Closed-form solution for design of lead-lag compensators. Int J Elect Eng Educ 41:172–180
Nassirharand A, Karimi H (2006) Nonlinear controller synthesis based on inverse describing function technique in the MATLAB environment. Adv Eng Softw 37:370–374
Nassirharand A, Mousavi Firdeh SR (2008) Design of nonlinear lead and/or lag compensators. Int J Control Automat Syst 6:394–400
Nassirharand A, Mousavi Firdeh SR (2009a) Combustion chamber pressure control based on mixed H_∞-describing function approach. Acta Astronaut 64:206–211
Nassirharand A, Mousavi Firdeh SR (2009b) Single-range controller design for a cruise missile. Aircr Eng Aerosp Technol 81:283–287
Nassirharand A, Mousavi Firdeh SR (2009c) Design of nonlinear H_∞ controllers using describing functions with application to servomechanism. Asian J Control 11:446–450
Nassirharand A, Patwardhan A (1988) A fast algorithm for obtaining low-order coprime fractional representations of SISO systems. Control Theory Adv Technol 4:277–281
Nassirharand A, Taylor JH (1990) Synthesis of linear PID controllers for nonlinear multivariable systems. In: Proceedings of the American control conference, San Diego, California, pp 2223–2228
Nassirharand A, Taylor JH (1991) Frequency domain modeling of nonlinear multivariable systems. Control Theory Adv Technol 7:201–214
Nassirharand A, Taylor JH, Reid KN (1988) Controller design for nonlinear systems based on simultaneous stabilization theory and describing function models. ASME Trans J Dyn Syst Meas Control 110:277–281
Nassirharand A, Karimi H, Dadfarnia M (2003) A new software tool for synthesis of linear PID controllers. Adv Eng Softw 34:551–557
Nelder JA, Mead R (1965) A simplex method for function minimization. J Comput 7:308–313
Nett CN, Jacobson CA, Balas MJ (1984) A connection between state-space and doubly coprime fractional representations. IEEE Trans Automat Control 29:268–280
Ogata K (1990) Modern control engineering, 2nd edn. Prentice Hall, Englewood Cliffs
Ozer MB, Ozguven HN, Royston TJ (2009) Identification of structural non-linearities using describing functions and the Sherman-Morrison method. Mech Syst Signal Process 23:30–34
Patwardhan A (1988) Compensator synthesis using factorization approach. Masters Thesis, Department of Mechanical Engineering, University of Kentucky, Lexington, Kentucky
Petrovic I, Baotic M, Perik N (2000) Model structure selection for nonlinear system identification using feedforward neural networks. In: Proceedings of IEEE-INNS-ENNS international joint conference on neural networks, Como, 1:1053

Prazenica RJ, Brenner MJ, Kurdila AJ et al. Nonlinear system identification for aero-elastic systems. Research and Engineering Education Facility, University of Florida, Shalimar Florida. http://www.reef.ufl.edu/nonlinear%20System%20Id%20fo..ro-ElasticB.htm

Pujara LR (1988) Computer-aided control systems design technique with applications to aircraft flying qualities. AIAA J Guid Contr Dynam 11:250–255

Reklaitis GV, Ravindran A, Ragsdell KM (1983) Engineering optimization. Wiley, New York

Rivera DE, Morari M, Skogestad S (1986) Internal mode control. PID Controller design. Ind Eng Chem Process Des 25:252–265

Rowland JR (1986) Linear control systems. Wiley, New York

Safonov MG, Limebeer DJN, Chiang RY (1989) Simplifying the H_{∞} theory via loop-shaping, matrix-pencil and descriptor concepts. Int J Control 50:2467–2488

Shi J, Sun HH (1990) Nonlinear system identification for cascaded block model: an application to electrode polarization impedance. IEEE Trans Biomed Eng 37:574–587

Skogestad S, Postlethwaite I (2005) Multivariable feedback control. Wiley, West Sussex

Slotine JJ, Li W (1991) Applied nonlinear control. Prentice Hall, Englewood Cliffs

Suzuki A, Hedrick JK (1985) Nonlinear controller design by an inverse random-input describing function methods. In: Proceedings of American control conference, Boston, MA, pp 1236–1241

Taylor JH (1983) A systematic nonlinear controller design approach based on quasilinear system models. In: Proceedings of American control conference, San Francisco, CA, pp 141–145

Taylor JH, O'Donnell JR (1990) Synthesis of nonlinear controllers with rate feedback via sinusoidal input describing function methods. American control conference, San Diego, pp 2217–2222

Taylor JH, Strobel KL (1985) Nonlinear control system design based on quasilinear system models. In: Proceedings of American control conference, Boston, MA, pp 1242–1247

Teh SH (2010) Design of multivariable dual-range linear controllers for nonlinear systems. BEng Thesis, Faculty of Engineering, The University of Nottingham, Semenyih, Selangor

Teh SH, Nassirharand A (2010a) Idle speed control using a describing function approach. Proc Inst Mech Eng Part D J Automob Eng 224:533–545

Teh SH, Nassirharand A (2010b) Software for design of linear controllers for nonlinear systems. In: Proceedings of international conference on advances in mechanical engineering, Shah Alam

Teh SH, Nassirharand A (2010c) Multivariable temperature control of a reactor. Proceedings of international conference on advances in mechanical engineering, Shah Alam

Ujile A (2010) MATLAB® software for the design of dual-range linear controllers with application to robotics. BEng Thesis, School of Electrical and Electronics Engineering, The University of Nottingham – Malaysia Campus, Semenyih, Selangor

Ujile A, Nassirharand A (2010) MATLAB software for design of SISO dual-range linear controllers. Int J Syst Control 1:149–155

Utkin V, Guldner J, Shi J (1999) Sliding mode control in electromechanical systems. CRC Press, Boca Raton

Vaerenbergh SV, Via J, Santamaria I (2007) Nonlinear system identification using a new sliding-window kernel RLS algorithm. J Commun 2:1–8

Vidyasagar M (1985) Control system synthesis: a factorization approach. MIT Press, Cambridge

Vidyasagar M, Viswanadham N (1982) Algebraic design techniques for reliable stabilization. IEEE Trans Automat Control 27:1085–1095

Wang L, Barnes TJD, Cluett WR (1995) New frequency-domain design method for PID controllers. IEE Proc: Control Theory Appl 142:265–271

Youla DC, Bongiorno JJ, Jabr HA (1976) Modern Wiener-Hopf design of optimal controllers – pt. 1: the single-input case. IEEE Trans Automat Control AC-21:3–14

Zhuang M, Atherton DP (1993) Automatic tuning of optimum PID controllers. IEE Proc Part D 140:216–224

Ziegler JG, Nichols NB (1942) Optimum settings for automatic controllers. ASME Trans J Dyn Syst Meas Control 62:759–768

Index

A. Nassirharand, *Computer-aided Nonlinear Control System Design: Using Describing Function Models*, DOI 10.1007/978-1-4471-2149-7,
© Springer-Verlag London Limited 2012

MIX
Papier aus verantwortungsvollen Quellen
Paper from responsible sources
FSC® C105338

If you have any concerns about our products,
you can contact us on
ProductSafety@springernature.com

In case Publisher is established outside the EU,
the EU authorized representative is:
Springer Nature Customer Service Center GmbH
Europaplatz 3, 69115 Heidelberg, Germany

Printed by Libri Plureos GmbH
in Hamburg, Germany